The Language of Science

'Professor Reeves transforms the achievements of two generations of creative scholarship in the language and rhetoric of science into a textbook that is fully accessible to under-graduates, while remaining informative for graduate students. She accomplishes this feat in a style that is patient without condescension, clear without oversimplification, and accurate without pedantry.'

Alan Gross, *University of Minnesota, USA*

'Thought-provoking tasks and commentary make this a highly interactive and informative 'read', dramatically and clearly illustrating how the use and construction of language can both shape and challenge our conceptions about the scientific enterprise.'

Anthony Blake, *University of Newcastle upon Tyne, UK*

The INTERTEXT series has been specifically designed to meet the needs of contemporary English Language Studies. *Working with Texts: A Core Introduction to Language Analysis* (second edition, 2001) is the foundation text, which is complemented by a range of 'satellite' titles. These provide students with hands-on practical experience of textual analysis through special topics, and can be used individually or in conjunction with *Working with Texts*.

The Language of Science:

◎   explores the goals of, and problems with, scientific discourse

◎   demonstrates the power and misuse of scientific discourse in the media

◎   examines the special qualities of scientific communication and rhetoric

◎   explores how science and popular culture interact

◎   is illustrated with a wide range of examples from the discourses of the AIDS epidemic to depression to global warming

◎   includes ideas for student enquiry projects and a glossary

Carol Reeves is Professor of English at Butler University, Indianapolis. She is an experienced teacher of rhetoric and science communication, and has published a number of articles on topics such as Language and AIDS.

Inter te**X**t

The Intertext series

The Routledge INTERTEXT series aims to develop readers' understanding of how texts work. It does this by showing some of the designs and patterns in the language from which they are made, by placing texts within the contexts in which they occur, and by exploring relationships between them.

The series consists of a foundation text, *Working with Texts: A Core Introduction to Language Analysis,* which looks at language aspects essential for the analysis of texts, and a range of satellite texts. These apply aspects of language to one particular topic area in more detail. They complement the core text and can also be used alone, providing the user has the foundation skills furnished by the core text.

Benefits of using this series:

◉ **Multi-disciplinary** – provides a foundation for the analysis of texts, supporting students who want to achieve a detailed focus on language.

◉ **Accessible** – no previous knowledge of language analysis is assumed, just an interest in language use.

◉ **Student-friendly** – contains activities relating to texts studied, commentaries after activities, highlighted key terms, suggestions for further reading and an index of terms.

◉ **Interactive** – offers a range of task-based activities for both class use and self-study.

◉ **Tried and tested** – written by a team of respected teachers and practitioners whose ideas and activities have been trialled independently.

The series editors:

Adrian Beard was until recently Head of English at Gosforth High School, and now works at the University of Newcastle upon Tyne. He is a Chief Examiner for AS and A Level English Literature. He has written and lectured extensively on the subjects of literature and language. His publications include *Texts and Contexts* (Routledge).

Angela Goddard is Head of Programme for Language and Human Communication at the University College of York St John, and is Chair of Examiners for A Level English Language. Her publications include *Researching Language*.

Core textbook:

Working with Texts: A Core Introduction to Language Analysis
(second edition, 2001)
Ronald Carter, Angela Goddard, Danuta Reah, Keith Sanger and
Maggie Bowring

Satellite titles:

Language and Gender
Angela Goddard and Lindsey Meân Patterson

Language Change
Adrian Beard

The Language of Advertising: Written Texts
(second edition, 2002)
Angela Goddard

The Language of Children
Julia Gillen

The Language of Comics
Mario Saraceni

The Language of Conversation
Francesca Pridham

The Language of Drama
Keith Sanger

The Language of Fiction
Keith Sanger

The Language of Humour
Alison Ross

The Language of ICT: Information and Communication Technology
Tim Shortis

The Language of Magazines
Linda McLoughlin

The Language of Newspapers
(second edition, 2002)
Danuta Reah

The Language of Poetry
John McRae

The Language of Politics
Adrian Beard

The Language of Speech and Writing
Sandra Cornbleet and Ronald Carter

The Language of Sport
Adrian Beard

The Language of Television
Jill Marshall and Angela Werndly

The Language of Work
Almut Koester

The Language of Websites
Mark Boardman

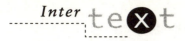

The Language of Science

©   Carol Reeves

Routledge
Taylor & Francis Group

LONDON AND NEW YORK

First published 2005
by Routledge
2 Park Square, Milton Park, Abingdon, Oxon OX14 4RN

Simultaneously published in the USA and Canada
by Routledge
270 Madison Ave, New York, NY 10016

Routledge is an imprint of the Taylor & Francis Group

Typeset in Stone Sans/Stone Serif by
Florence Production Ltd, Stoodleigh, Devon
Printed and bound in Great Britain by
TJ International Ltd, Padstow, Cornwall

British Library Cataloguing in Publication Data
A catalogue record for this book is available from the
British Library

Library of Congress Cataloging in Publication Data
Reeves, Carol
  The language of science/Carol Reeves
      p. cm. – (Intertext)
  Includes bibliographical references
  1. Communication in science.   2. Science – Language.   3. Technical
  writing.   I. Title.   II. Series: Intertext (London, England)
  Q223.R44 2005
  808'.0665–dc22   2005004872

ISBN 0–415–34635–5 (hbk)
ISBN 0–415–34636–3 (pbk)

contents

introduction

For many of us, science is where we turn for facts, for answers to questions about ourselves and the world around us. Those answers, we assume, are **objective** and untainted by politics and **bias**, unlike explanations that do not come from careful experiment and measure. So when we hear in the news that 'scientists report' a new finding or when 'scientists say' that we should redirect our lifestyle, we usually listen. Such an authoritative arena ought to be closely examined and better understood.

One way to understand science is to examine its language. This book is about the language used within science to explain, argue and characterize phenomena as well as the language used outside science to explain and report its findings to all of us. It is important to know both aspects of scientific language, the esoteric and the popular, in order to think more carefully about how science impacts our lives.

The Index of terms gives a short explanation of special terms, or terms used in a specialized way, with the page number where it is first used. Each term is **emboldened** on its first mention only.

WHAT IS SCIENCE?

When most people think of science, they think of individuals – Albert Einstein, Charles Darwin, Dr Frankenstein from the movies, Carl Sagan and Stephen Hawkins. Because scientific discoveries or revolutionary ideas are often attached to individuals, we often assume that science is made up of solitary 'nerds' who go about their business without collaboration with others. But this is a misconception created by our exposure to popular narratives that dramatize the individual experience.

Science, like all other realms of human activity, is inherently social. Scientists, like others at work, must collaborate in order to be productive, and in order to collaborate, they must use oral and written language. They work together in the laboratory, sharing **data**, arguing over expla-nations, exchanging ideas. They e-mail their colleagues at other labora-tories and talk at conferences. Writing, in any laboratory, is just as important as the actual experiments: careful laboratory notes must be taken, findings must be interpreted and reported, papers must be drafted, read, rewritten and then sent to a journal where they are then read, sent back, rewritten and eventually published to be read and discussed by audiences. Audience reactions – whether they agree or disagree or whether they find a scientific paper relevant – depend as much upon *how* a paper was written as upon the information it reports. No data set can stand alone. Scientists must explain their interpretations of data and make reasonable arguments in defence of their claims about their data. How well they accomplish this important professional task depends upon their command of scientific language and **rhetoric**.

LANGUAGE AND SCIENCE

Language, then, is just as crucial to science as it is to poetry and politics, religion and business, or any other arena of knowledge. But as we know from experience with these areas, the relationship between language used to represent some actual reality (an experience, a feeling, a belief, a new car) and the actual reality itself is slippery. That is, the words we choose in these areas to represent anything may be motivated by our desires – to persuade, to entertain, to enlighten. We accept that language use in poetry and politics, religion and business is biased and arbitrary.

But what about language use in science? And what about scientific language on our televisions and in our newspapers and magazines? Is the language of science, both specialized and popularized, just as motivated and biased and slippery as the language used in other areas? This book will provide some answers to these questions. But a simple answer is yes and no. Yes, the language of science, as language, is always imperfect. There is no perfect, absolutely accurate way to describe what lies under the microscope or under our skin or out in space.

The scientific community has been battling the conundrums of language for centuries. Good scientists are good linguists insofar as they are vigilant in their reading and writing, keenly aware of the slippery nature of language and careful of the language choices they make. So, no, scientific language is not as emotional as political or religious language or as overtly persuasive as the language we use in other realms, but it is, nonetheless, still a system made by human beings who cannot entirely control the effects their language choices have on others.

ANALYSING SCIENTIFIC LANGUAGE

This book introduces you to several ways you can analyse and, thus, better understand scientific language and, hopefully, the scientific enterprise. Whether you are pursuing a career in science or in the humanities, communications or business or just planning to raise a family one day, you will need to understand scientific language. Curiosity may move you to learn more about Mars or the search for extraterrestrial life or illness may lead you to medical reports about a particular disease. One day, your own child may ask why. At the very least, you should be able to read scientific literature on your own with confidence. And you should have a healthy radar that will detect the political, economic and social motivations that exploit science for profit, influence and power.

So what do we analyse when we analyse scientific language?

We analyse the goals of scientific language and **terminology**. Why is scientific language different from poetic language? What are some problems with scientific terminology? These questions will be addressed in Unit one.

We analyse the use of **metaphor** *in science*. Why do scientists use metaphors? Do metaphors work in science the same way they work in ordinary communication? Do metaphors influence the way scientists think? How might metaphors cause problems for scientists? These questions will be addressed in Unit two.

We analyse what makes scientific writing sound so scientific. Why is the writing in a scientific research report so different from the writing in a novel? When we examine features that all areas of science share, we are examining **scientific discourse**. What are the grammatical and organizational features that scientists employ in order to be viewed as scientists rather than as English teachers or rock musicians or self-help gurus? Scientific discourse is the subject of Unit three.

We analyse patterns in scientific discourse that signal speculation as well as certainty. What patterns of language do scientists use when they have a hunch that something is true but they don't yet have the **evidence**, that is, when they speculate? What patterns of language do scientists use when they present **scientific facts**? What are the patterns of language that characterize the movement of an idea from speculation to fact? These questions will be addressed in Unit four.

We analyse how scientists persuade each other, how they employ **scientific rhetoric** *to argue and support their claims*. How do scientists persuade each other without resorting to the strategies of modern advertising and politics? Scientists hold one another to a high standard of evidence and reasoning when they make claims about their data;

3

however, as we know, data may be interpreted in more than one way. For all sorts of reasons, scientists must understand scientific rhetoric, the art of persuading a scientific audience that a claim is valid and viable given the available evidence. Rhetoric is the subject of Unit five.

We analyse how the language and discourse of science interact with and influence communities outside science and how the culture of those outside communities influences science. How and when does science take the lead in shaping our understanding of an issue? How and when do religious, political or economic concerns prevail over science in shaping our understanding of an issue? These questions will be addressed in Unit six.

We analyse the translations and translators of science. Who commonly translates scientific findings for the rest of us? What motivations lie behind a translation of science? When are translations misleading and misguided? When are they effective? Most of us are so familiar with the use of science in advertising or the explanation of scientific discoveries in the news that we may not pay much attention. But we owe it to ourselves and our communities to pay attention. The translation of science is the subject of Unit seven.

Unit one

Language

Like all of us, scientists must find words to describe, explain and name the concepts, objects and processes they examine. Their explanations and descriptions, their names for objects and concepts, become very important in communicating ideas and generating knowledge.

However, no explanation and no term can capture entirely all the features of any phenomenon. Scientists are interested in finding out what is generally true under most conditions, so their language typically captures what is most generally observable and predictable about phenomena. For example, here is an excerpt from the most recent clinical definition of Major Depressive Disorder used by psychologists and physicians for diagnosis:

Major Depressive Episode

Episode Features
The essential feature of a Major Depressive Episode is a period of at least 2 weeks during which there is either depressed mood or the loss of interest or pleasure in nearly all activities. In children and adolescents, the mood may be irritable rather than sad. The individual must also experience at least four additional symptoms drawn from a list that includes changes in appetite or weight, sleep, and psychomotor activity; decreased energy; feelings of

worthlessness or guilt; difficulty thinking, concentrating, or making decisions; or recurrent thoughts of death or suicidal ideation, plans, or attempts. To count toward a Major Depressive Episode, a symptom must either be newly present or must have clearly worsened compared with the person's pre-episode status. The symptoms must persist for most of the day, nearly every day, for at least 2 consecutive weeks. The episode must be accompanied by clinically significant distress or impairment in social, occupational, or other important areas of functioning. For some individuals with milder episodes, functioning may appear to be normal but requires markedly increased effort.

The mood in a Major Depressive Episode is often described by the person as depressed, sad, hopeless, discouraged, or 'down in the dumps' (Criterion A1). In some cases, sadness may be denied at first, but may subsequently be elicited by interview (e.g., by pointing out that the individual looks as if he or she is about to cry). In some individuals who complain of feeling 'blah,' having no feelings, or feeling anxious, the presence of a depressed mood can be inferred from the person's facial expression and demeanor. Some individuals emphasize somatic complaints (e.g., bodily aches and pains) rather than reporting feelings of sadness. Many individuals report or exhibit increased irritability (e.g., persistent anger, a tendency to respond to events with angry outbursts or blaming others, or an exaggerated sense of frustration over minor matters). In children and adolescents, an irritable or cranky mood may develop rather than a sad or dejected mood.

(*Diagnostic and Statistical Manual*: 349)

Those who study and treat mental illnesses must be able to share a descriptive or explanatory language that captures those experiences most often reported by those with mental illnesses. Without a clear physical sign or laboratory test for depression, psychiatric professionals must rely on patients' descriptions of their mental or emotional state to diagnose problems. Without a reliable standard description of the most common symptoms of depression, psychiatric professionals would be unable to diagnose patients or communicate clearly with each other. In using the *Diagnostic and Statistical Manual*'s (*DSM*) definition of depression, professionals try to match their patients' language with the standard language of depression in the *DSM*. In the 'Major Depressive Episode' passage, notice how physicians are directed in how to 'read' their patients' language and translate it into the language of depression.

A patient might not describe herself or himself as 'sad' but the physician may 'elicit' that response 'through interview'. Depression can be 'inferred from the person's facial expression and demeanour' even when he or she 'complain[s] of feeling blah'. Notice also the use of phrases such as 'In some cases', 'In some individuals' and 'many individuals'. The clinical picture provided is attempting to account for the inevitable differences in human beings' experience of mood disorders. Due to patients' cultural background, gender and age, as well as physical condition, they may experience and describe their depression in different ways. Yet to ensure greater diagnostic accuracy, the clinical language of depression must capture only those experiences that most patients report. Scientific language must capture the general, the usual, rather than the individual and the unique.

SCIENTIFIC VS POETIC LANGUAGE

One way of understanding how scientific language works is to contrast it with the way poetic language works. Scientists and poets share the desire to find the most precise and effective language to convey experience. Both rely on careful observation and keen perception to interpret all types of 'data', whether in the laboratory or in the everyday world of human relations. Both poets and scientists admire elegance, whether in the cadence and melody of a line or in the capaciousness of an equation.

Yet, you know very well that a poem and a scientific report seem to have been produced by beings inhabiting different planets. Many of the striking differences you find are the result of cultural and professional traditions that have developed over time due to the separation of science and poetry as two distinct communities. The scientific community wants to establish what is true or constant most of the time under most conditions, what is generally applicable to various contexts. The poetic or literary community wants to establish what is singular, **subjective**, ambiguous and mutable. Poets hope that after reading their poems, readers will be able to appreciate the singular, individual experience that may or may not be repeatable. Scientists hope that after reading their research reports that readers will be able to repeat the **methods** or apply the data in new situations.

Activity

1   Read Poem 258 by the American poet Emily Dickinson (see Text 1).
Paying close attention to the italicized words, try to sum up what you
think Dickinson is saying about depression or despair. Do you think
that Emily Dickinson would have believed that the sort of depression
she describes here ought to be treated?

2   Re-read the *DSM* definition of depressive disorder given earlier
(pp. 5–6). What main differences do you see in the two texts?

Text 1

There's a *certain Slant of light,*
Winter Afternoons –
That *oppresses*, like the Heft
Of Cathedral Tunes –

Heavenly Hurt, it gives us –
We can *find no scar*,
But *internal difference*,
Where the *Meanings, are* –

None may teach it – Any –
'Tis the *Seal Despair* –
An imperial affliction
Sent us of the Air –

When it comes, the Landscape listens –
Shadows – hold their breath –
When it goes, 'tis like the Distance
On the look of Death –

While both texts are about depression, they each say something very different about it. In the scientific text, depression is an illness to be treated while in the poem, it is as an altered or higher state of being, a 'Heavenly Hurt' that leads to 'internal' 'Meanings' and an 'imperial affliction' whose coming and going can be compared to the quality of light across a winter landscape. The scientific text is intended to convey what the psychiatric community agrees is the most generally true description of the symptoms of depression while the poem conveys only one person's subjective experience, which may or may not be true for any other person's experience. The language in both texts is precise and descriptive, but each text conveys only part of the reality of depression. *DSM* emphasizes those parts of depression that are most commonly experienced and thus most easily contribute to a diagnosis of disorder. The poem emphasizes the less common metaphysical parts of depression that are experienced individually and may resist diagnosis or treatment, that may, in fact, be signs of strength rather than weakness.

SCIENTIFIC LANGUAGE SHOULD BE FREE OF BIAS AND EMOTION

The founders of the Royal Society of London, the UK National Academy of Science founded in 1660, stressed that scientific language should carefully describe nature. In 1667, Thomas Spratt insisted on language that captured 'observation of Nature in its particulars [with] a long forbearing of speculation at first' (in Sutton, 1994: 55). In fact, the motto of the Royal Society is 'Nullius in Verba', which translated literally means 'nothing in words'. However, members of the Royal Society did not mean that nothing could be put into words. Their intention was to promote greater care in the use of language and greater scrutiny of knowledge claims that had not been thoroughly tested. They urged scepticism of claims stated by authoritative individuals or persuaded through bombastic rhetoric. They aimed for a language that was lean and as close to the details of nature as possible.

The Royal Society's emphasis on careful, descriptive language has been misunderstood to mean that language and facts are in opposition.

The general public tends to assume that scientists *report facts* rather than *interpret* data or even *develop ideas*. We assume that the language of science is only a simple, descriptive system, thus much less important than practical experience. This is why science education so often reduces language to the reading of textbooks, the memorization of textbook facts, and the following of instructions for laboratory experiments. Students rarely learn that in scientific knowledge, there is no escaping human involvement with words. Without language, there would be no facts.

Still, you have probably already noticed that the language of science is very different from the language of political campaigns or the language of advertising or poetry. The goals of careful description of nature are still very important. Scientists try to find the most efficient and objective language to characterize the appearance, function or composition of phenomena.

The goal of scientific language is to be as free as possible from **connotations** that reflect or create cultural biases and emotional attachments. But even when scientists are very careful to avoid connotations, the very language they choose can grow out of or become associated with particular attitudes and prejudices. Even more important, scientific language used to describe what scientists study can emerge from cultural assumptions and biases that scientists may not recognize as such.

A good example of how cultural connotations grow out of or influence scientific language can be found in the language employed by psychologists to describe homosexuality.

Until 1971, the American psychological community considered homosexuality as a mental health problem. Severe pressure from a society unwilling to acknowledge or accept homosexuality drove homosexuals to psychotherapy for treatment. Because there is no diagnosable physical factor in homosexuality, psychology became the segment of the medical community destined to define and describe it as an illness. The language developed to describe and define homosexuality both reflected and perpetuated the assumption that it was an illness, an assumption held by psychiatrists, the general public and homosexuals themselves.

Yet, views about homosexuality in the society of psychotherapists did change as did their language. These changes resulted from insights gained from their patients as well as shifting social views. These changes may be traced in the definitions and descriptions of homosexuality in the *DSM*.

Below are clinical descriptions from successive editions of the *DSM* used by physicians and psychologists to diagnose mental illnesses and personality disorders:

1950s

000–x60 Sociopathic Personality Disorder

Individuals to be placed in this category are ill primarily in terms of society and of conformity with the prevailing cultural milieu, and not only in terms of personal discomfort and relations with other individuals. However, sociopathic reactions are very often symptomatic of severe underlying personality disorder, neurosis, or psychosis, or occur as the result of organic brain injury or disease. Before a definitive diagnosis in this group is employed, strict attention must be paid to the possibility of the presence of a more primary personality disturbance; such underlying disturbance will be diagnosed when recognized. Reactions will be differentiated as defined below.

000–x63 Sexual Deviation

The term includes most of the cases formerly classed as 'psychopathic personality with *pathologic sexuality*'. The diagnosis will specify the type of the pathologic behavior, *such as homosexuality* . . .

(Diagnostic and Statistical Manual, 1st edn)

1960s

302 Sexual Deviations

This category is for individuals whose sexual interests are directed primarily toward objects other than people of the opposite sex, toward sexual acts not usually associated with coitus, or toward coitus performed under *bizarre circumstances as in necrophilia, pedophilia, sexual sadism, and fetishism.* Even though many find their practices distasteful, they remain unable to substitute normal sexual behavior for them. This diagnosis is not appropriate for individuals who perform deviant sexual acts because normal sexual objects are not available to them.

302.0 Homosexuality [a type of sexual deviation]

(Diagnostic and Statistical Manual of
Mental Disorders, 2nd edn)

11

1970s

302.00 Ego-dystonic Homosexuality

The essential features are a desire to acquire or increase hetero-sexual arousal, so that heterosexual relationships can be initiated or maintained, and a sustained pattern of overt homosexual arousal that the individual explicitly states has been unwanted and a persistent source of distress.

This category is reserved for those homosexuals for whom changing sexual orientations may be a brief, temporary manifestation of an individual's difficulty in adjusting to a new awareness of his or her homosexual impulses. Individuals with this disturbance may have either no or very weak heterosexual arousal. Typically there is a history of unsuccessful attempts at initiating or sustaining heterosexual relationships. In some cases no attempt has been made to initiate a heterosexual relationship because of the expectation of lack of sexual responsiveness. In other cases, the individual has been able to have short-lived heterosexual relationships, but complains that the heterosexual impulses are too weak to sustain such relationships. When the disorder is present in an adult, usually there is a strong desire to have children and family life.

Generally, individuals with this disorder have had homosexual relationships, but often the physical satisfaction is accompanied by emotional upset because of strong negative feelings regarding homosexuality. In some cases the negative feelings are so strong that the homosexual arousal has been confined to fantasy.

(*Diagnostic and Statistical Manual III*, 3rd edn)

Current

Gender Identity Disorder can be distinguished from simple non-conformity to *stereotypical sex-role behavior* by the extent and pervasiveness of the cross-gender wishes, interests, and activities. This disorder is not meant to describe a child's nonconformity to stereotypic sex-role behavior as, for example, in 'tomboyishness' in girls or 'sissyish' behavior in boys. Rather, it represents a profound disturbance of the individual's sense of identity with regard to maleness or femaleness. Behavior in children that merely does not fit the cultural stereotype of masculinity or femininity should not be given the diagnosis unless the full syndrome is present, including marked distress or impairment.

(*Diagnostic and Statistical Manual IV-TR*)

Notice the changes in the way homosexuality has been defined for those diagnosing mental illnesses and personality disorders. In the 1950s, homosexuality was 'pathologic' and labelled a 'sexual deviation', which was categorized as a 'personality disorder'. Such specialized terms were likely selected for their denotative meanings. However, as we all are well aware, these terms carry negative connotations that likely reflected the psychological community's misunderstanding of and disgust with homosexuality. Their language spread into popular usage as homosexuals were labelled 'deviates' and 'sociopaths', which further spread unfortunate prejudices.

In 1971, the psychological community agreed that homo-sexuality is not, in and of itself, an illness that requires treatment. Yet, in the 3rd edition of the *DSM*, homosexuals are still viewed as seeking relief from their homosexual urges. Their illness is characterized by low heterosexual response and shameful homosexual desires. The latest version of the *DSM* does not contain the word 'homosexual' at all.

The goal of objective and denotative scientific language is often difficult to achieve. No matter how carefully language is selected to describe the observable features of phenomena, bias and cultural meanings can always subvert the goal of objectivity. Humans are fallible and so is the language we employ. Scientists cannot entirely rule out their biases at the beginning of their investigations. The development of scientific knowledge rests on accounts that eventually become accepted by the community as reasonable interpretations of what happens in nature. The language of description can harbour unfortun-ate biases that, hopefully, will be eliminated in the movement toward knowledge.

SCIENTIFIC LANGUAGE AND THE
AIDS EPIDEMIC

As we have said before, no linguistic utterance is entirely representative of reality. In making choices about how to describe or explain an idea, writers or speakers are making linguistic selections based on what they see, what they have experienced and what they believe. Sight, experience and belief are all limited. Scientists may never see the whole object or they may have limited laboratory experience with it, or their beliefs about what they think they will find in their investigations may create a mindset that prevents them from finding something else entirely.

Scientists, whether they are an astrophysicist searching the stars or a doctor seeing a new disease for the first time, must find language to report what they have seen. But their description is always going to be partial, especially at first, because their experience with the phenomenon is limited. Yet, their limited description will be read by others and will shape their understanding of the phenomenon.

The first physicians in the United States to treat what is now known as **AIDS** had to alert the medical community to a new disease that could become an epidemic. Their descriptions had to capture those aspects of the disease that other physicians could recognize in their own patients. Unfortunately, the first patient population in the United States were social pariahs – homosexuals and injection drug users. For many physicians, their first AIDS patients were homosexuals.

The first medical descriptions of AIDS necessarily captured what seemed to be the most observable aspects of the disease – that the patients were homosexual and or IV drug users, that they were dying of infections or cancers that normal immune function would prevent. While these first descriptions of AIDS were clinical and legitimate, they created conceptions that directed and limited both the medical community's and the popular community's understanding of this new disease. Based on the first medical reports, the newspapers labelled the new disease 'Gay Cancer' and 'Gay-related Immune Deficiency'.

Text 2 shows how this process works. On the far left is the reality facing those first to treat AIDS. On the far right is the 'reality' that first emerged early in the epidemic after several reports of a new disease among gay men.

14

Activity

What effects do you think the conceptions about AIDS that resulted from the language used to report the new disease had on average people and their understanding of the new disease? What effects do you think the early conceptions of AIDS had on scientific research?

Text 2

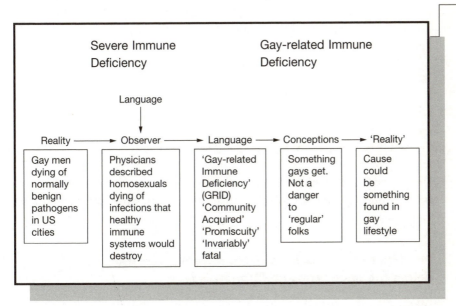

Commentary

Conceptions embedded in a discourse that has derived from individual reports and observations are not necessarily reality. Any set of terms used to describe a phenomenon does capture some aspects of that phenomenon, but it usually does not capture the entire picture. **Clinical discourse** about AIDS early in the epidemic captured what the first clinicians saw – patients who were in many ways social pariahs dying of unusual infections.

The first physicians to treat and describe AIDS in the United States in 1981 called the new syndrome 'Gay-related Immune Deficiency' or GRID. Because most of the first AIDS patients were gay, physicians used the term 'gay-related' to indicate what they wrongly assumed was an objective feature of the new disease. Unfortunately, 'gay-related', once circulated in the media, conveyed cultural meanings that allowed many people to assume that only homosexuals were at risk. Eventually, the new disease was called 'Acquired Immunodeficiency Syndrome' (AIDS). More denotative and objective than GRID, the new term included the word 'acquired' to indicate that the immune deficiency was not the result of a precondition such as malnutrition or drugs to suppress the immune system such as chemotherapy. However, 'acquired' carried an additional meaning – that of behaviour leading to infection. It is true that human beings 'acquire' the disease; however, the word 'acquired' carried a subtle cultural connotation that placed blame on the infected for 'acquiring' the disease in the first place. The current term, 'Human Immunodeficiency Virus or **HIV**' erases all such connotations.

PROBLEMS WITH SCIENTIFIC TERMINOLOGY

Despite the care scientists take to develop precise, reliable and meaningful terminology, many problems arise due to confusing, misleading and vague terms.

Problem I: When scientists from different subfields define the same terms in different ways

American evolutionary biologist Stephen Jay Gould (1988) provides an example of confusing terminology that led to a disastrous outcome. As Professor Gould demonstrates, confusion about the meaning of a technical term, **'homology'** may have led one researcher to incorrectly assume that a baboon heart could be transplanted into a human infant.

On 26 October 1984, Dr Leonard L. Bailey of the Loma Linda University School of Medicine transplanted the heart of a baboon into the chest of Baby Fae, an infant born with a congenital heart malformation. The child at first responded well, but her body eventually rejected the baboon heart, and she died twenty days after the transplant.

What would have led a scientist to take such a hopeless measure? Professor Gould argues that scientific language influenced Dr Bailey's thinking and his actions. Gould explains that Dr Bailey may have felt comfortable transplanting a baboon heart into Baby Fae because of confusion about the meaning of the word 'homology'. Two fields, evolutionary biology and biochemistry had different meanings for the same word.

According to Professor Gould, evolutionary biology distinguishes between two very different types of similarity between organisms, one based on **genetics** and the other based on similar adaptation to a common function:

1   *Homology*: similarities between organisms due to descent from a common ancestor. For example, arm and finger bones in mammals are homologous.

2   *Analogy*: similarities arising as separate evolutionary adaptations to common functions. For example, the wings of bats, insects and birds.

Comparative biochemistry, the molecular study of phylogeny, searches for the criteria that establish whether organisms are similar due to homology or **analogy**. However, unlike evolutionary biologists who employed both terms to distinguish both types of similarity, biochemists began to use the word 'homology' for both types of similarity as long as there was a certain percentage of matching **DNA** bases in molecular comparisons. For a common ancestor, maybe you could have 80 per cent matching DNA bases, but less than that percentage could be caused by analogy. However, the word 'homology' was used to indicate similarities between organisms due to a descent from a common ancestor (true homology) as well as similarities arising as separate evolutionary adaptations to common functions (analogy). Our similarity to baboons is due to adaptation to common function, not descent from a common ancestor, thus baboons cannot donate organs to humans.

Here is how Dr Bailey explained why he used a baboon heart rather than a human or even a chimpanzee (close to us genetically) heart:

> Chimpanzees seem to share greater homology with humans, but are virtually unavailable as organ donors. Baboons are generally less concordant with man than are chimpanzees, but it may be feasible to narrow the 'disparity gap' by a careful selection. ... Some degree of homology between baboon and human lymphocyte antigens must exist.
>
> (Gould, 1988: 31)

Professor Gould insists that Bailey uses the word homology 'in the improper sense of measured overall concordance, rather than the evolutionary meaning of similarity due to descent from a common ancestor' (ibid.) and because of this, Bailey 'is led to the false idea that he might find overlap between the closest baboon and the furthest chimp – so that a particular baboon might be an acceptable candidate, even though chimps are closer to people on average' (ibid.).

Because two very different meanings of similarity had been combined into one term, Dr Bailey assumed that any degree of similarity between the human infant and a baboon would be sufficient for transplantation.

Problem 2: When terms are vague, they can mean different things to different people

The term Prion is now the name for the agent that causes diseases such as Bovine Spongiform Encephalopathy (BSE) and the human forms of the disease such as Crutzfeld Jacob's Disease. The term was introduced by American scientist Dr Stanley Prusiner in 1981. Prion originally stood for 'Proteinaceous Infectious Particle' because Dr Prusiner and his laboratory team had discovered that a protein played an important role in the disease. Dr Prusiner, who eventually won the Nobel Prize in 1997, believed that these diseases are caused primarily by infectious protein particles that contain no nucleic acid to aid replication. However, the question of whether Prions actually work without any help from nucleic acid has never been completely settled.

The problem, as Reeves points out (2002), with the term Prion is that for many years, between 1982 and the late 1990s, it was employed to represent two very different things:

1   An infectious protein particle without nucleic acid.

2   An infectious protein particle that may contain nucleic acid.

Writers of scientific research reports who did not believe that Prions acted without nucleic acid employed the term Prion, wrongly assuming that their meaning of the term was stable. Their readers could read the term and apply whichever definition they assumed to be true. The end result was confusion and misunderstanding (Reeves, 2002).

Problem 3: Inappropriate scientific terms

Melvin Konner (2001) explains that terms are inappropriate if they are based on theories that cannot or have not been tested. If these terms catch on, they spread false theories. Sigmund Freud coined the term 'anal fixation' to describe his **theory** that children pass through an anal stage of development. While such a stage has not been scientifically validated, the term 'anal' has seeped into popular usage as indicating obsession with orderliness or with hoarding. But the original Freudian association between the potty and orderly habits is unfounded.

Another inappropriate term, pointed out by Konner, that has gained popular acceptance is B.F. Skinner's term for the brain as a 'black box', as an organ whose inner workings were determined by environment rather than genetics. Skinner believed that the route to understanding what happens inside the 'black box' was to observe outward behaviour. Despite growing knowledge of the genetic bases of learning and other cognitive behaviour, the term 'black box' persists in the discourse of psychology.

SUMMARY

In this chapter, we have explored the goals of scientific language and terminology. Because most of us are exposed to scientific language through scientific textbooks, we assume that scientific language is impenetrable and unchanging, like the language of some exclusive club, interpretable by members only. But like any language, scientific language is a human product and as such, it is penetrable and always changing in response to changing circumstances. Unlike the language we use and hear every day, scientific language aims to be denotative and objective, to capture as faithfully as possible the features of natural phenomena without bias or emotional and cultural connotation. However, language always manages to resist these goals. Like all of us, scientists are influenced by their culture and emotions, and their language reveals these influences and thus can never be entirely unbiased and objective.

Extension

1   Many writers have written about their experience with depression. Read one of the following works or find your own. What is the author's individual expression of mental illness? How well does their experience fit into the diagnostic definition of depression provided at the beginning of this chapter?

William Styron, *Darkness Visible*

Sylvia Plath, *The Bell Jar*

Elizabeth Wurtzel, *Prozac Nation: Young and Depressed in America: A Memoir*

2   Interview a scientist, one of your teachers or someone you know. Ask the following questions:

   a   Describe any problems with language that you have noticed in your own field.

   b   Are there any confusing or problematic terms used in your field?

FURTHER READING

Darian, Steven, *Understanding the Language of Science*, Austin: The University of Texas Press, 2003.

Gould, Stephen Jay, 'The Heart of Terminology', *Natural History* 2 (1988): 24–31.

Heilbron, J.L., 'Coming to Terms', *Nature* 415 (2002): 585.

Keller, Evelyn Fox, 'Language in Action', in Matthias Dorries, ed., *Experimenting in Tongues*, Stanford, CA: Stanford University Press, 2002: 76–88.

Konner, Melvin, 'Bad Words', *Nature* 411 (2001): 743.

Raad, B.L., 'Modern Trends in Scientific Terminology: Morphology and Metaphor', *American Speech* 64.2 (1989): 128–36.

Reeves, Carol, 'An Orthodox Heresy: Scientific Rhetoric and the Science of Prions', *Scientific Communication* 24.1 (2002): 98–122.

Sutton, Clive, '"Nullius in verba" and "nihil in verbis": Public Understanding of the Role of Language in Science', *British Journal of the History of Science* 27 (1994): 55–64.

SOURCE

Diagnostic Statistical Manual IV-TR, Washington, DC: American Psychiatric Association, 2000.

Metaphor in science

We would be unable to think or communicate without metaphors. They enable us to express the unknown or unseen via the known and seen. They help us understand what we can't touch, feel or see through what we can touch, feel and see. For example, we apply our experience of the concrete world and physical space in metaphors that describe our feelings and our thoughts. We say we are feeling 'up' or 'down', that we have something 'on' our minds. Some people are 'insiders' and some are 'outsiders'. Some are 'small-minded' while others have 'big hearts'.

In many ways, all language is metaphoric if we think of metaphor as simply the application or mapping of one world of experience or domain onto another in order to describe or explain that domain. If you were asked to describe this book, or any textbook, you might use a metaphor that joins your experience in another domain with the experience of reading this book:

'It's a chore.'
'It's illuminating.'
'It's eye-opening.'
'It's a sleeper.'

We often fail to realize how much we rely on metaphors, and thus fail to understand how much they influence us. Not only do they help us express what we can't sense directly, they also shape our attitudes and opinions. When we use metaphors, whether consciously or unconsciously, we are mapping some aspect of one domain of experience onto another. The link between these two domains is only partial. Even as that link allows us to express or explain some aspects of the 'target' domain, it hides other aspects. If we never critically think about our metaphors, we are doomed to live by their partial truth.

For example, we can identify an interesting metaphoric exchange between the domains of medicine and war in the following metaphors:

We are the frontline in the battle against AIDS. Eventually, we will defeat the enemy with the ultimate weapon – a vaccine.

The surgical strikes hit their intended targets with no collateral damage. The operation was a clean sweep of the territory.

The metaphor of battle is ubiquitous in medicine. Doctors and scientists are depicted as valiant soldiers in the battle against the enemy diseases. The battle metaphor is helpful in that it creates an optimistic attitude that eventually the war will be won and our side will win. Yet the battle metaphor is not always helpful. While modern medicine has made enormous advances in the prevention and treatment of many diseases, it will never conquer the enemy. Unrealistic expectations of medical victory blind us to the realities of medical failures around the world – the emergence of new infectious diseases, the lack of quality medical care in the third world, the profit motive that drives much of our drug research.

The medical and sanitary metaphors in military language also create confidence in the precision and effectiveness of modern military weapons and personnel. But mostly, medical metaphors obfuscate the ugly realities of war, allowing us to feel comfortable about dropping bombs.

As you move through your day, jot down statements you read or overhear that contain metaphors. Then, analyse a few of the metaphors you found. What serves as the informing domain and what domain is it used to explain? What underlying attitudes do these metaphors support?

(Note: there is no commentary for this activity.)

METAPHORS IN SCIENCE

Metaphors are inescapable in science just as they are in everyday language. As human beings who must often draw from various domains of experience in order to make sense of new domains of experience, scientists are no different from the rest of us. And in many ways, science is no different from other realms of human activity in its reliance on metaphors. In fact, many linguists and scientists are coming to view metaphor as central to scientific thought and practice. Chemist Theodore Brown (2003: 15) writes in his book on science and metaphor:

> [Metaphor] lies at the heart of what we think of as creative science: the interactive coupling between model, theory, and observation that characterizes the formulation and testing of hypotheses and theories. None of the scientist's brilliant ideas for new experiments, no inspired interpretations of observations, nor any communications of those ideas and results to others occur without the use of metaphor.

As with other areas of human community and language, in science, metaphor works on several levels:

- Metaphors can serve as models of processes or objects that scientists cannot see.

- Metaphors can become theories that explain and predict the behaviour or action of processes and objects.

- And metaphors can help scientists explain and communicate complex ideas to non-scientists and science students.

We must keep in mind, however, that metaphors may hide as much as they reveal. No metaphoric link is an absolute match. As we saw earlier, medicine is not really war, and war is not really medicine, but metaphors that map each of these areas onto the other create prevailing ideas, good and bad, that may support or disable progress. The same is true for metaphors in science.

Activity

Text 3 contains descriptions of how **retroviruses** cause our immune system cells to die. See if you can identify the metaphor used to explain this process. What gets emphasized about the 'primary' system (retroviral action) as a result of the 'secondary' (the familiar domain)?

Text 3

Once infection occurs, and the T4 cells are activated, instead of yielding 1,000 progeny [as in a healthy system] the infected T cell proliferates into a stunted clone with perhaps as few as 10 members. When those 10 reach the blood stream and are stimulated by antigen, they begin producing virus and die.

(Reeves, 1992: 334)

[the cell's] death may depend on an interaction between the viral envelope and the cell membrane. Perhaps that interaction . . . punches a hole in the membrane. Because the virus buds in a mass of particles, the cell cannot repair the holes as fast as they are made; its contents leak out and it dies.

(Reeves, 1992: 334)

Commentary

In the first passage, we have the battle metaphor that pervades the language of **immunology**. The infected cells do not 'yield' a healthy number of 'progeny' but instead become 'stunted clones' with few 'members' that go into the bloodstream and produce virus. The primary system is the intercellular activity that occurs once the virus enters the immune cells. The secondary system is the familiar ideas of reproduction, activation, enervation and death. Do cells have 'progeny'? Do they have 'members' like a team? Do they 'die' in the way we think of death or do they simply dissolve? These metaphors emphasize those aspects of intercellular and viral activity that are similar to what we understand as reproduction, activation, enervation and death, but they also may suppress very important aspects of the 'primary' (intercellular and viral activity) that may not be similar to our understanding of these processes.

The second description that our scientists called 'for the moment only a model' employs a metaphor from our everyday experience. Terms such as 'punch out' and 'leaks out' suggest puncturing a bag full of water. As the authors admit, this metaphor serves as both a way to explain this process and as a **hypothesis**.

METAPHORS AS MODELS

Let us explore how metaphors are models of processes and objects we cannot see. Theodore Brown (2003: 23) explains that we often become so accustomed to a model that we forget it is a metaphor. He provides as an example the common representations of the methane molecule:

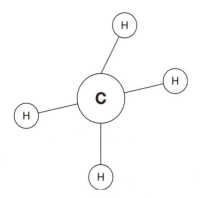

Brown explains that this model contains two metaphoric conceptions: that atoms are spheres and that chemical bonds are rigid rods between atoms. Experimental work in chemistry has produced certain facts that are captured in this model:

1   There are four hydrogen atoms and one carbon atom in a methane molecule.

2   The hydrogen atoms all have the same relationship with the carbon atom. They are not chemically bonded to one another.

3   Atoms are spherical.

4   There is a characteristic distance between the carbon and each hydrogen atom.

So the spheres and the rigid rods in the model fit the observations made in experiments. However, it is still a metaphor because the model cannot be seen as a literal representation. Brown continues:

> The ball-and-stick model accounts for important aspects of the methane molecule, but it is inconsistent with others. Experimental evidence indicates that atoms are not like hard spheres but more like rubber balls that grow softer at their outer edges. A hard rubber ball on the inside with a Nerf ball-like exterior would serve as an appropriate metaphor.
>
> (ibid.: 24)

He adds that models like this one 'have become so commonplace in scientific explanation, and they can be so beautiful, that it is easy to succumb to the idea that they are literal descriptions' (ibid.: 24).

METAPHORS AS THEORIES

In addition to serving as illustration, metaphors can also become theories if they can be used to make predictions and plan experiments as well as explain phenomena. In 1609, astronomer Johanes Kepler wrote that his goal in writing *Astronomia Nova* was 'to show that the celestial machine is to be likened not to a divine organism but rather to a clockwork' (quoted in Johnson-Sheehan, 1998: 175). Kepler reveals his preference for one of two main theories in conflict during his lifetime

– the theory of 'nature as an organism' and the theory of 'nature as a machine'. These expressions, 'nature as an organism' and 'nature as a machine' are **'root metaphors'**, which are simple, elegant fundamental ideas that are easily applied to a variety of natural phenomena. Here are some of the meanings of each metaphor:

Machine:
  input, output, parts, tools, efficiency, productivity, goal.

Organism:
  self-preserving, adaptability, individuality, reproduction.

Activity

See if you can apply both of these metaphors to the phenomena in Text 4. What aspects of each phenomenon does each metaphor explain? What ideas do the metaphors generate about each phenomenon?

Text 4

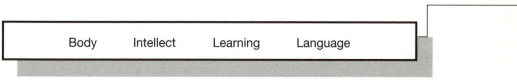

| Body | Intellect | Learning | Language |

(Note: there is no commentary for this activity.)

Scientific theories often begin as metaphors that have been selected by scientists to help them explain their ideas. For example, some protein molecules are labelled 'chaperones', because they are a class of proteins that help other proteins achieve their appropriate structure and to assemble correctly with other proteins. The first scientists to use the term 'chaperone' were applying its familiar meanings – 'protector' or 'usherer' – to express their idea that some molecular mechanism was helping crucial interactions in the cell. But as Brown (2003: 153) explains, the chaperone concept quickly became a theory that took on 'a life of its own' in guiding new hypotheses about the role of chaperone functions in cellular biology.

Activity

Text 5 is an example of scientists employing the chaperone concept in their thought processes. What new metaphor do they bring into their hypothesis?

Text 5

> The dictionary defines triage as 'sorting and allocation of treatment to Patients.' The patients in this case are cellular proteins. The first level of triage must be identification of the proteins that are damaged and require treatment. . . . Once damaged proteins have been identified, a second level of decision must be made: Can the patient be saved? Chaperones or chaperone components . . . should have the first opportunity to correct misfolded proteins. Hopeless cases in which structural damage cannot be repaired need to be degraded.
>
> (Brown, 2003: 155)

Commentary

The domain of the hospital serves as a source of the metaphor used to explain how chaperone proteins repair damaged proteins.

METAPHORS AS THEORIES ABOUT ATOMIC STRUCTURE

Theories about the structure of the atom are metaphoric. As Brown explains, the metaphors used by scientists to model atoms became theories guiding the design of experiments that would demonstrate whether those metaphorical theories were correct.

The plum pudding metaphor

One early model of atomic structure, developed by British scientist J.J. Thomson in the late 1800s, was known as the 'plum pudding' model of the atom (ibid.: 79). This model was proposed as a way of explaining the distribution of positive and negative charges within the atom. In this model, the atom consisted of a cloud of positive charge kept in balance by electrons, the cloud being the pudding and the electrons being the plums. This **plum pudding model** guided the design experiments used to test its correctness. In these experiments, beams of X-rays were broken up so that scientists could then further examine the distribution of positive and negative charges. These experiments showed that there were fewer electrons, and thus fewer negative charges, than had been assumed. According to what physicists believed at the time, this situation was counter-intuitive; it simply did not make sense. With fewer negative charges, what would keep the positive charges in check? How could the positive and negative charges within the atom be kept in balance if there weren't enough electrons? It seemed that the plum pudding model would need to be revised. According to Professor Brown, J.J. Thomson decided that he would stay with his model because it conformed to his experimental results, and because he 'was prepared to find that some of the familiar laws of physics would ultimately prove not to be applicable' (ibid.: 80). Professor Brown explains that the plum pudding model, though imperfect, is an example of how metaphors serve as 'tools of discovery' by 'point[ing] the way toward new experiments even as their apparent failings call attention to the need for improvement or replacement' (ibid.: 80).

Following the plum pudding model, Ernest Rutherford, designed experiments in 1906 at Manchester University to study the scattering of alpha particles. Earlier he had identified the three forms of radioactive emissions, the alpha, beta and gamma emissions. In his scattering experiments, Rutherford passed an alpha beam through foil. Since alpha particles were known to be larger and heavier than electrons, they were not expected to be very scattered when they hit the foil. This also fits in with the plum pudding model – heavier 'plums' would more likely 'sink' into the batter than be deflected by it. The experiment's result was surprising: they found that a small number of alpha particles had been deflected at large angles, which indicated that collisions had occurred between the heavier alpha particles and something else. This would seem to be evidence against the plum pudding model, yet Rutherford did not immediately abandon the metaphor. It took him a year to rethink the old theory of atomic structure and replace it with the idea of the nuclear atom. To explain why some of the heavier alpha particles were scattered, Rutherford developed the **nuclear model** of the atom in which there is a small, massive charged particle, the nucleus, in the atom that is strong enough to deflect those alpha particles that happen to come in contact with it.

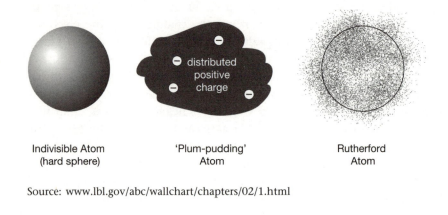

Indivisible Atom   'Plum-pudding'   Rutherford
(hard sphere)   Atom   Atom

Source: www.lbl.gov/abc/wallchart/chapters/02/1.html

METAPHORS USED TO TEACH SCIENCE

Non-scientists interested in science and science students often benefit from scientists' use of metaphor to explain complex ideas and processes. Communicative metaphors may have no theoretical use for scientists but they can bring science into our familiar domains. Sometimes, metaphors drawn from social life may not be familiar to popular audiences. In the passage below, the image of the locomotive is used as a metaphor for an assembly of proteins that assist DNA replication:

> We could now turn to phiX174 and begin to explain the molecular operations of the multiprotein assembly (called a primosome) that starts a DNA chain. The image of a locomotive seemed helpful for a time in accounting for primosome actions. The engine, protein n' . . . powered by ATP energy, is a helicase; it unzippers the DNA duplex and is equipped with a cowcatcher to remove SSB (single stranded binding protein) in its path. Another protein, dnaB, is both helicase and engineer, using ATP to locate or shape a section of DNA track upon which primase will find it possible

to lay down a short stretch of RNA, which will then attract DNA polymerase to start a DNA chain. The primosome is translocated on DNA in only one direction, the one that keeps it at the advancing fork of a replication chromosome.

(Kornberg, 1989: 19)

Kornberg admits that the locomotive image 'seemed helpful for a time', but may no longer serve explanation, especially if readers are not familiar with the term 'cowcatcher'.

Activity

Read the explanation in Text 6 of how HIV (the Human Immunodeficiency Virus known to cause AIDS) enters our cells. Identify the metaphors used to explain this process.

Text 6

HIV gains entry to its target cell with a pair of protein molecules called gp120 and gp41, which are attached like a hair to the virus's outer membrane, or 'skin'. Part of this molecular strand of 'hair' is attracted to three other kinds of molecules attached to the surfaces of HIV's target cells, which the strand fits like a key in a lock. One of the target-cell molecules is a protein called CD4, which occurs on two cells of the immune system: helper T cells and macrophages. (Helper T cells recognize foreign substances and regulate the immune response to them. Macrophages ingest foreign particles and eliminate them from the body.) The second target-cell molecule for HIV is a receptor known as CCR5, which normally binds to a small protein secreted by white blood cells when they encounter foreign materials. The third receptor, known as the fusion domain, is the target of gp41. When the viral protein gp120 binds to both a CD4 molecule and a chemokine receptor, and gp41 binds to the fusion domain, a 'door' to the target cell swings open. The viral membrane fuses with the cell membrane, and the virus enters the cell.

(Rajan, 2004: 40)

Commentary

Here Dr Rajan uses metaphors familiar to us as well as metaphors common in the language of immunology to illustrate how HIV enters the host cell; hair, keys and locks, and doors allow him to illustrate a process we will likely never observe. But he also uses ingrained metaphors that have long been part of the language of immunology. Like the language of genetics, the language of immunology is replete with metaphors. The T lymphocytes have been described as 'conductors' of the immune system 'orchestra', while battle metaphors characterize the relationships between foreign or enemy invaders and our 'helper' and 'killer' lymphocytes.

> 'Helper T Cells recognize foreign substances'.
> 'Macrophages ingest foreign particles'.
> 'The second target-cell ... which normally binds to a small protein'.
> 'The third receptor, known as the fusion domain'.

We have several metaphors here that are perhaps not recognized as such by those who are familiar with this language. Clearly, the domain of ordinary experience which includes 'helping', 'recognizing' and 'ingesting' or eating is mapped onto experimental observations to create this explanatory language. Yet, T cells may not be 'helping' but merely 'reacting', and they may not 'recognize' substances as 'foreign' but simply be genetically programmed to respond to particular signals. HIV may not actually 'target' certain cells as much as it is attracted by them. These metaphors have simply been borrowed from other domains to express what has been observed in experiments. There is no absolutely literal description that captures entirely the truth of these processes.

Activity

Find a current textbook used to teach any scientific discipline. Browse through the text to find metaphors used to teach concepts to students. What familiar domains are being mapped onto the process or object being taught?

(Note: there is no commentary for this activity.)

THE LANGUAGE OF GENETICS

It is impossible to talk about genetics without the metaphor of language – of reading, writing and communicating. Human language, as grammar, alphabet, writing, reading and communicating, has served as an important domain for explaining what happens inside our genes. The complex interactions between the processes whereby our DNA replicates and shapes us are captured in the language or communication metaphors – reading and writing, translating, transcribing, coding and decoding. For example, the genetic 'message' is carried by the 'messenger' RNA to the ribosomes where it is 'translated' into protein. In fact, DNA itself is often conceptualized 'as a work of literature, a great historical text. But the metaphor of a chemical text is more than a vision: DNA is . . . similar in function . . . to the letters of a book . . .' (Pollack, 1989: 4). Indeed the general themes of reading and writing pervade discussions of DNA:

> DNA is like a book of instructions in each cell. The alphabet used to create the book is simple enough: A, T, G and C. But how is the alphabet arranged into the sentences (genes) that become expressed as proteins? How does a cell skip through a book, reading only those genes that will provide specific proteins at specific times?
>
> (Darian, 2003: 95)

Evelyn Fox Keller (2002) believes that the language or the reading/ writing metaphor has not only helped us explain or illustrate genetics but has also influenced certain patterns of thinking among microbiologists and geneticists. She suggests that the reading and writing metaphors have allowed scientists to think of themselves as 'readers' and 'rewriters' of the **human genome**.

On the website for the United States National Institutes of Health Human Genome Project, the following statement by the director, Francis Collins, is used to explain the importance of the map of the human genome:

> It's a history book – a narrative of the journey of our species through time. It's a shop manual, with an incredibly detailed blueprint for building every human cell. And it's a transformative textbook of medicine, with insights that will give health care providers immense new powers to treat, prevent and cure disease.

Like all metaphors, the 'Book of Life' metaphor emphasizes some ideas over others. In this optimistic statement, our power not just to read these books of the human genome but to 'rewrite' them in order to prevent and cure disease is emphasized. The possibility that we won't be good readers or that our 'rewriting' will destroy the 'books' of life is left out of this highly enthusiastic pronouncement.

It is possible that one day, we will see the 'Book of Life' metaphor in genetics to be quaintly simplistic. The structure of language and the systems of interpersonal communication has served in helping us think about how DNA, genes and proteins interact. But the language metaphor carries some assumptions that may not be true when applied to our genes. When we communicate, we know how to recognize when a message is being sent, to whom and by whom, and we know this because we share a language that makes this knowledge possible. However, in genetics, the movement of messages from a sender to a receiver is not at all so simple or so discoverable.

Language is more than a tool for describing and reporting observations and experience. Language both limits and enables our ability to think about and understand our world. Even in Science, with its standards of objectivity and demonstration, the language used to describe, explain, illustrate and model phenomena can engender a way of thinking about the phenomenon that may or may not be entirely correct.

Activity

Identify the metaphors in the passages in Text 7. What is their purpose? To illustrate or model? To teach? Are they intended for other scientists and thus serve as a way to theorize or are they intended for nonscientists and thus serve to explain?

Text 7

1

Let us conclude this chapter with a case study of how the immune system helps you survive attack . . .

All this time your body had been struggling against an unseen enemy. During a walk, one of your feet had picked up some soil bacteria. And when the tack broke through your skin, it carried several thousand bacteria cells inside it . . . their [the bacteria's] metabolic products were interfering with your own cell functions. If unchecked, the invasion would have threatened your life.

If this had been your first exposure to the bacterial process, few B and T cells would have been around to respond to the call. . . . But when you were a child, your body did fight off this invader and still carries the vestiges of the struggle – memory cells. As inflammation progressed, B and T cells were also leaving the bloodstream. Most were specific for other antigens and did not take part in the battle. But memory cells locked onto the antigens and became activated. For the first two days the bacteria appeared to be winning. They were reproducing faster than the phagocytes . . . were destroying them. By the third day, antibody production peaked and the tide of battle turned. For two weeks more, antibody production will continue until the invaders are wiped out. After the response draws to a close, memory cells will go on circulating, prepared for some future struggle.

(Darian, 2003: 99)

2

But in fact, a breed, like a dialect of language, can hardly be said to have a distinct origin. A man preserves and breeds from an individual with some slight deviation of structure, or takes more care than usual in matching his best animals, and thus improves them . . .

(Darwin, 1859: 40)

Commentary

1 The battle metaphor, as we have seen, is a standard metaphor in immunology, the study of immune response. This metaphor provides a good theory of the interaction of our immune cells and viruses and bacteria entering the body. In many ways, the process is like a battle involving defensive and attacking forces. In the passage, the author uses the metaphor dramatically to teach his audience about antibody response.

2 In this passage Charles Darwin uses the metaphor of language dialect in order to argue that animal breeds, like dialects (and by implication, other species, including humans) have evolved over time, due to some type of selection of strengths, rather than to a single creative event. Throughout his *Origin of the Species*, Darwin uses analogies, such as animal breeding, and metaphors of language, in this case dialects, to persuade his readers that life on earth has evolved over time as a result of natural selection.

SUMMARY

Just like the rest of us, scientists cannot escape metaphor. Either they employ metaphors intentionally, to explain and illustrate natural phenomena or they use them unconsciously because some metaphors are so firmly entrenched that they go unnoticed. Whether they are used consciously or unconsciously, metaphors affect all of us by subtly shaping and limiting our view of the world.

FURTHER READING

Brown, Theodore, *Making Truth: Metaphor in Science*, Urbana and Chicago: University of Illinois Press, 2003.

Darian, Steven, *Understanding the Language of Science*, Austin: University of Texas Press, 2003.

Darwin, Charles, *On the Origin of Species*, London: John Murray, 1859.

Hesse, Mary, *Models and Analogies in Science*, London: Sheed & Ward, 1963.

Johnson-Sheehan, Richard, 'Metaphor in the Rhetoric of Scientific Discourse', in John T. Battalio, ed., *Essays in the Study of Scientific Discourse*, Stamford, CT: Ablex Publishing, 1998: 167–80.

Keller, E.F., *Refiguring Life: Metaphors of Twentieth Century Biology*, NY: Columbia University Press, 1995.

Keller, Evelyn Fox, 'Language in Action: Genes and the Metaphor of Reading', in Matthias Dorries, ed., *Experimenting in Tongues: Studies in Science and Language*, Stanford: Stanford University Press, 2002: 76–88.

Lakoff, George and Mark Johnson, *Metaphors We Live By*, Chicago: University of Chicago Press, 1980.

Pollack, R., *Signs of Life: The Language and Meanings of DNA*, London: Viking, 1989.

Reeves, Carol, 'Owning a Virus: The Rhetoric of Scientific Discovery Accounts', *Rhetoric Review* 10.2 (Spring 1992): 321–36.

SOURCES

Kornberg, A., 'Never a Dull Enzyme', *Annual Review of Biochemistry* 58 (1989): 19.

Rajan, T.V. 'Fighting HIV with HIV', *Natural History* February (2004): 38–44.

three

The grammar of science

We have explored the terminology and linguistic metaphors that not only communicate but also contribute to scientific thinking and knowledge. Another important element of scientific language is grammar – the ordering of words within sentences and the structure of sentences. If you have ever tried to read a scientific article in a specialized journal, you may have given up in exasperation. Not only is the terminology esoteric, the formulas incomprehensible, and the visuals inexplicable, but the sentences themselves are simply impossible to decipher. Sometimes scientists are simply bad writers. The great evolutionary biologist and essayist, Stephen Jay Gould (2003: 132), insists that there is no reason why scientific prose should be so unreadable:

> Because we have cut ourselves off from scholars in the humanities who pay closer attention to modes of communication, we have spun our own self-referential wheels and developed artificial standards and rules of writing that virtually guarantee the unreadability of scientific articles outside the clubhouse.

But those who have studied the development and practice of scientific prose style argue that, bad writing aside, scientific prose style serves scientific thinking and practice in several important ways. They assert that some identifiable features of scientific grammar or prose style

37

accommodate theory building as well as the need for efficient and economical communication.

THE THEORY OF GRAMMATICAL METAPHOR

Linguist M.A.K. Halliday (1998) has a theory about how and why scientific writing differs from other types of writing. Halliday teaches us that scientific discourse exploits a capacity in our language that all of us use every day without realizing it. This he calls '**grammatical metaphor**'.

Grammatical metaphor is the process whereby we think theoretically. The process goes like this:

Step One: We have an experience.

Step Two: We 'construe' our experience in language. That is, we tell ourselves or someone else about our experience.

Let us pretend we had this experience:

Before the performance of Second City at Clowes Hall Saturday night, we *drank* free wine at the subscriber's lounge without *showing* our tickets to prove we were subscribers.

According to Halliday, the grammar of our language not only allows us to represent our experience but to 'reconstrue' it, shape and reshape it for different purposes and different contexts. The way our language allows this reconstruing of experience lies at the heart of how we theorize, not just in science but also in our everyday lives.

This '**reconstrual**' of our experiences through language involves shifting from what Halliday calls the '**congruent**' to the 'technical'. The congruent could be called the first step of our making sense of experience. The congruent style of expression captures the experience before we have begun to reconstrue it. Above is a 'construal' of an experience. This expression might be called 'congruent'.

Step Three: We 'reconstrue' our experience by rethinking and restating it as a theory about what sort of experiences we can expect in the future.

As Halliday points out, our grammar allows us to 'reconstrue' our experience by changing verbs and adjectives into nouns. This is called '**nominalization**', and it moves our expression toward the theoretical. We can rethink our experience and our rethinking is captured in a shift in the parts of speech. Notice how we can create 'noun groups' from verbs in a 'reconstrual' of our experience:

Drinking free wine before a performance is a perk for those who hold season tickets.

Showing your ticket is not needed.

We now have two theories based on grammatical reconstrual. One is that if we purchase season tickets next year, we can have free wine before each performance. But we may also theorize that we might be able to drink free wine without being season ticket holders if we simply go to the lounge and look like we know what we're doing. The point is that as we move from verbs to nouns, whether just in our consciousness or in our conversation, we are moving from experience to theory about future experience.

Grammatical metaphor is Halliday's term for the shifting of one grammatical category to another. This is not the same as *a regular or lexical metaphor* which implies an abstract relationship between two things that belong to different domains or categories.

Unlike **lexical metaphor**, implying an abstract relationship between two things that belong to different categories, grammatical metaphor refigures and reverses the status of actions, attributes and things. A chain of **linguistic transformation** occurs in which actions or attributes (expressed as verbs or adjectives) become things (expressed as nouns).

Here is another illustration:

Congruent construction:

Sara	puts money	into the meter	so	she	won't get	ticketed.
Figure	process	circumstance	relator	figure	process	process

Metaphoric 'reconstrual':

Putting money into the meter is *prevention* against ticketing. (Both of these grammatical metaphors are shifts in rank and status.)

Puts (verb/process) → Putting (noun/thing)

won't get (verb/process) → Prevention (noun/thing)

Activity

Text 8 contains statements offered by my students as examples of the congruent expression of experience. Restate these statements by shifting the verbs or attributes to nouns so that you create expressions that are more theoretical.

Text 8

I parked in the visitors' parking even though I am a student but didn't get a ticket.

Mike plays cricket, and he's very clever.

I fell asleep in a large class, but I still made a good grade.

Commentary

Notice how easily we can draw inferences, or theories, from these statements. We may theorize that if on one day, we park in the visitors' parking without being ticketed, we may be able to do so again: parking in the visitors' parking is allowed. Or we might reconstrue the second statement as 'Playing cricket requires clever athletes'. Or if we get away with sleeping in class without failing our class, we may imagine that 'sleeping in large classes is fine, even good for you'. The point is that the expectations or theories that we build from single experiences are supported by the very grammar in our language, so that it may be impossible to think theoretically without language as we know it.

Activity

Text 9 shows examples of the reconstrued or technical style. See if you can shift the statements back into the congruent.

Text 9

Skimming over your reading assignments is enough to get by in any class.

Liberalism is the expected outcome if you major in political science here.

Grade inflation is what happens when students evaluate their professors.

Commentary

These are examples of statements that result from grammatical reconstrual. Each statement hints at an original experience upon which the theory is grounded. For example, if one student simply skims a reading assignment and manages to do well in the class, she may theorize that skimming is the way to get through long reading assignments. Or if a student experiences one liberal political science professor, she may assume the entire group of professors are liberals. Or one professor who fears bad evaluations and inflates the grades to make students happy may assume that all professors do the same and thus that there is a cause–effect relation between student evaluations and grade inflation. You can see the danger in the ease with which our language allows us to theorize from single experiences. We make false, even dangerous, assumptions when we assume that if X is true in one situation, it will always be true.

GRAMMATICAL METAPHOR AND SCIENTIFIC THEORY

The process of grammatical reconstrual lies at the heart of how scientific theories are developed. This process may evolve over an extended period of time in which scientists engage in experimental experiences that they construe and reconstrue in language. Sometimes we may observe this process artificially, in the texts they write to communicate their findings and ideas. These texts may contain grammatical metaphors that convey or mirror the reconstrual processes in the scientist's thinking. But they may also support the arguments the writers are making. Halliday gives the following example:

> If electrons weren't absolutely *indistinguishable*, two hydrogen atoms would form a much more weakly bound molecule than they actually do. The absolute *indistinguishability* of the electrons in the two atoms gives rise to an 'extra' attractive force between them.
> (1998: 202; emphasis mine)

This example contains a grammatical shift from attribute (indistinguishable) to 'thing' (indistinguishability). Notice here how the writer smoothly shifts the adjective 'indistinguishable' to the noun 'indistinguishability'. A quality becomes a thing that becomes part of a theory of how atoms interact.

When writers change verbs or adjectives into nouns, they can 'create' objects out of processes, qualities and attributes. Grammatical metaphor allows communicators to accomplish several important tasks. In the example below, the congruent statement describes an action while in the reconstrual, the action has become a noun and its status as an objective phenomenon is solidified:

Congruent construction:

The normal protein *misfolds* into an abnormal form.

Technical reconstrual: (from verb to noun).

Misfolding of protein is the initial event in a chain of infection.

In explaining how protein can cause disease, scientists studying Bovine Spongiform Encephalopathy (BSE) and other similar diseases began with the observation that proteins can fold the wrong way. When proteins fold the wrong way, they can create the form of protein that is found in infected tissues. Eventually, this observation is reconstrued into theoretical statements in which the action 'folding' becomes a thing, the noun 'misfolding'. The grammatical transition from verb to noun allows the building of a theory of disease causation in which misfolding is the initial event in disease.

GRAMMATICAL METAPHOR AND
SCIENTIFIC TERMINOLOGY

Scientists studying BSE and other related diseases now use the word 'Prion' to stand for the cause of the diseases. This term conveys the idea that protein can cause disease without the help of nucleic acid.

Compacting

The term 'Prion' is actually the result of the *compacting* of different grammatical elements into one:

Congruent description, what 'Prion' originally stood for:

'Proteinaceous	Infectious	Particle'
Adj/quality	adj	entity

Grammatical shift:

Prion (compacts the descriptors or qualities into a thing, Prion)

Compacting allowed the controversial idea of an infectious protein to be 'hidden' inside one term. Scientists began to use the term 'Prion' even when they disagreed with the idea of an infectious protein. In a way, we can say that the term 'Prion' behaved like an infectious agent, 'infecting' the field with the theory that protein particles can cause disease without nucleic acid.

Distilling

The widespread use of the term 'Prion' is also the result of another grammatical metaphor: *distilling*. **Distilling** happens when a compacted term gets repeated and reconstrued as different grammatical elements. In our example, the term 'Prion' is shifted into new categories so that even more scientists employed the term:

Prionics   Prion Science   Prion Protein   Prion Protein Gene

With each distillation, 'Prion' becomes a new thing, an entity that participates in the generation and promotion of a theory of disease.

GRAMMATICAL METAPHOR AND
SCIENTIFIC ARGUMENTS

The two statements below appeared in a 1982 review essay in *Science* on the cause of Scrapie (a fatal disease in sheep that is similar to the disease most of us know as Mad Cow Disease or BSE). Notice the verb phrase in the first statement. Then look what happens to that verb phrase in the second statement that appeared later in the paper.

1   'Investigators *have so far been unable to find* a nucleic acid associated with disease in infected tissue' (Prusiner, 1982: 137).

2   'The lack of nucleic acid associated with disease in infected tissue is reason enough to consider alternative processes involved in disease causation' (ibid.: 144).

The verb phrase 'have so far been unable to find' in the first statement is shifted to a nominalization, 'the lack of nucleic acid'. Action – an inability to find – becomes a thing – the lack of – which becomes the basis for theoretical language. The writer of the theory, Stanley Prusiner, wanted to promote what is now called the 'protein-only' theory in which protein can replicate and infect tissue without nucleic acid. Grammatical metaphor allowed him to initiate and then to continue to promote this new theory of infectious disease.

WHAT GRAMMATICAL TRANSFORMATIONS REVEAL
ABOUT SCIENTIFIC EXPERIENCE

1   'Experience' can be mainly linguistic. All of the grammatical shifts described earlier occurred in a scientific field that had no clear answer to the question of what causes the diseases. Yet the language helped promote the idea that proteins could be infectious.

2   The movement from 'experience' to theory can begin with the movement from verb to noun, from 'unable to find' to 'the lack of'. Scientists knew that just because no nucleic acid had been found that that did not prove nucleic acid was not there. Very small nucleic acids might never be detected or it could be a virino with a

protein coat that prevents detection. Yet the grammatical metaphor encouraged a process of theory-building that rested upon the 'lack of' nucleic acid rather than the 'inability' to detect it.

3   The 'rhetorical package', that theoretical statement about the 'lack' pointing to a 'possibility', enacts then social networks among scientists who may stop looking for a nucleic acid and start studying proteins associated with the disease.

Halliday (2003: 187) says that 'the way things are is the way our grammar tells us that they are'.

But the good news is that by the same means that are used to construe and reconstrue experience, we can 'challenge the form of the reconstrual', that is, it is possible to challenge the meanings conveyed. Experience can be reconstructed in a different light.

GRAMMATICAL METAPHOR AND ACADEMIC WRITING

You may have noticed features of scientific discourse in many of your college textbooks. All branches of the social and natural sciences have adapted them. You might say that the features we have covered so far constitute a 'master' academic discourse that all disciplines with a scientific focus have adopted. Part of what is required to become a good student is to master this 'master discourse', to learn to read it and write it. When you try to understand your reading assignments, you are often called upon to translate technical discourse back into 'your own words', so to speak, into the congruent. The first step in that direction is to learn what that discourse is and why it is so difficult to understand.

SUMMARY

In this unit we have learned more about how language and thought are related. The grammar in our language allows us to 'play' around with our experiences as we put them into words. We might capture our experience in verbs, as actions, but very quickly we begin to form new statements that promote a different type of thinking, not just the kind that allows us to remember and record a recent experience but the kind of thinking that prepares us for the future. Scientists also generate theories in basically the same way although a good scientist would insist on repeated experience and replication before much confidence can be placed in a theory. Grammatical metaphor, the shifting from verbs to nouns in the expression of experience, is necessary to scientific thought and argument.

Extension

1   Go to one of your textbooks from the social or natural sciences. Pick out one sentence that you think is the result of grammatical metaphor (where actions or attributes have been shifted into nouns). Rewrite the sentence and eliminate nominalizations.

2   Write down a 'congruent' sentence that describes something you did or something that happened. Then transform this statement into a technical sentence like earlier examples by shifting grammatical categories. What theories about the future have you developed?

FURTHER READING

Gould, Stephen Jay, *The Hedgehog, the Fox, and the Magister's Pox: Mending the Gap between Science and the Humanities*, New York: Three Rivers Press, 2003.

Halliday, M.A.K., 'Things and Relations: Regrammaticizing Experience as Technical Knowledge', in J.R. Martin and Robert Veel, eds, *Reading Science: Critical and Functional Perspectives on Discourses of Science*, London and New York: Routledge, 1998.

SOURCE

Prusiner, Stanley B., 'Novel Proteinaceous Particles Cause Scrapie', *Science* 216 (1982): 136–44.

Discourse and facts

As we know from *Working with Texts* (238, 291–2), the term 'discourse' refers to the system of rules for language use that evolve in a community, whether by conscious choice or through cultural and other forces. 'Discourse' also refers to the patterns of language that can be identified as bound to a particular community and context. We can talk of 'scientific discourse' as the general language of science, the patterns of rule-governed language used among scientists.

All language is rule-governed. That is, when we speak or write, we must follow rules guiding pronunciation, word order and idiom as well as more specialized rules for communicating in a specialized community. In specialized communities, such as science, rules or conventions evolve over time, in response to new pressures and needs. Sometimes journal editors and writers make conscious decisions to change or add new rules for written communication that will ensure greater efficiency and clarity, but mostly these linguistic or discursive rules develop over time as a result of cultural norms and internal and external pressures. For example, the experimental article has shortened, become more technical and lost the narrative or 'storytelling' voice of its earliest forms.

Activity

Examine the two excerpts in Texts 10 and 11. What differences do you see between them? Examine the use of pronouns, active and passive voice. In what other ways do the two passages differ?

Text 10

The first paragraph of a paper from *Science*, I.2 (16 February 1883): Ira Remsen, 'Influence of Magnetism on Chemical Action'

More than a year ago I gave an account of some experiments which I had performed with the object of determining whether magnetism exerts any influence on chemical action. I succeeded in getting what appears to me to be strong evidence in favor of the view that magnetism does, at least in one case, exert a marked influence on chemic action. The principal experiment upon which this conclusion is based may be briefly described here. A vessel made of thin iron (ferrotype-plated were used) was placed on the poles of a magnet, and a solution of sulphate of copper poured into it. Instead of getting a uniform deposit of copper on the bottom of the vessel, the metal was deposited in distinctly marked lines, the direction of which was at right angles to the lines of magnetic force. Further, directly over the poles, the deposit was uniform; and this uniform deposit was bounded by a band of no deposit, from one-sixteenth to one-eighth of an inch in width.

Text 11

From *Inorganic Chemistry*, 40.22 (2001): 5581–4: Tosha M. Barclay, Robin G. Hicks, Martin T. Lemaire and Laurence K. Thompson, 'Synthesis, Structure, and Magnetism of Bimetallic Manganese or Nickel Complexes of a Bridging Verdazyl Radical'

Introduction

The design and construction of new magnetic materials from molecular components is a major contemporary theme in materials research. Among the many approaches to molecule-based magnets, hybrid materials comprised of paramagnetic metal ions and stable radicals offer several advantages. Direct metal-radical exchange interactions are possible, and the use of bridging radical-based ligands allows for the creation of extended metal-radical structures with cooperative magnetic properties. In the latter context, most work in this area has been carried out with bridging nitroxide radicals and diradicals. Complexes containing coordinated radical anions, for example, cyanocarbons such as TCNE or TCNQ or the semiquinones, have also received attention. However, the paucity of high Tc magnets in these systems highlights the continued need to explore alternative metal-radical assemblies.

We and others have recently demonstrated that judiciously substituted verdazyl radicals chelate to metals with structural features reminiscent of chelating oligopyridines. Thus, mononuclear Ni(II) and Mn(II) complexes of pyridine-substituted verdazyl 1 and related derivatives have been shown to possess strong metal-verdazyl magnetic exchange interactions, while bis(verdazyl) 2 has been incorporated into one-dimensional chains with Cu(I) halides. However, to create metal-radical assemblies with macroscopic magnetic ordering, there is a need to explore the efficacy of paramagnetic metal ions in conjunction with bridging verdazyls, in particular metal-radical exchange phenomena in bridged systems. As a first step toward creating extended arrays of transition metals and bridging verdazyls, we present herein the synthesis and characterization of two model binuclear compounds containing the bridging ligand 1,5-dimethyl-3-(4,6-dimethyl-2-pyrimidinyl)-6-oxoverdazyl radical, 3, a structural mimic of 2,2′-bipyrimidine, 4.

Commentary

Several changes in scientific discourse have occurred in the years since our first example was published. The single author is now rare since complicated laboratory work requires the collaboration of several specialists with training in specific techniques. The collaborative enterprise also leads to the loss of the **first-person pronouns** and active verbs seen in the earlier text. In the later piece, the plural first-person pronoun 'We' signals collaboration and the passive voice creates a more objective tone in contrast to the more personal and subjective tone of the earlier text. The *organization* of the samples is also very different. While the entire papers are not reproduced here, if you were to read them, you would find that the older text follows a narrative progression in which our scientist simply tells the story of his experiment and what he found. The modern text follows the conventions of arrangement of the **experimental report** that have solidified over the years. The modern experimental report is usually divided into four clear sections: introduction, methods, results and discussion.

These changes over the years in how experimental findings are reported result from a combination of forces. One is economic; a shorter, more concise paper costs a journal less money to publish and allows room for other reports. Another is cultural; the solitary storyteller or experimenter does not chime with the modern reality of collaboration as the source of scientific knowledge. In fact, the passive voice verb constructions in modern experimental papers support a cultural norm that displaces human agency, as if humans simply record, uncover, see a waiting world rather than manipulate that world to fit their presuppositions. So the experimental article is an artifice that displays science as an objective enterprise; it serves to promote and represent an objectivity that may be intended but never fully achieved, if only because scientists are human beings who must use human language to report their experiences.

THE EXPERIMENTAL REPORT IN SCIENCE

One rule-governed activity in science is the presentation of the results of scientific research. The form and style of the experimental report in science has solidified into a recognizable and repeatable genre. When scientists write papers to be published in their professional journals, they follow patterns and rules governing organization, diction and **register** as well as graphic and mathematical representation. The experimental report contains the following parts, all having their own complex rules about what can and cannot be said:

1   An *abstract* summarizes the report's main findings.

2   An *introduction* provides the background knowledge or general consensus in the field about what is known and unknown about the topic. It is also where writers justify their research and state their *research hypothesis*, which is the working idea that they either support or falsify in their experiments.

3   A *methods* section explains how the research was conducted.

4   A *findings* section *summarizes* the main results.

5   A *discussion* section offers an interpretation of the data, providing arguments in support of that interpretation.

Generally, the report follows a predictable pattern. The introduction moves from the known – what scientists agree are the facts surrounding a topic – to the unknown – what scientists need to better understand. Introductions also move from the background knowledge to a description of what the current research hopes to establish, their research purpose. The writers also explain their hypothesis, their working theory of what their research would reveal.

Here we will examine parts of a scientific research report in astronomy:

Activity

Read the abstract of this report in Text 12. What main findings are summarized?

Text 12

Abstract

In preparation for the advent of the Allen Telescope Array, the SETI Institute has the need to greatly expand its former list of ~2000 targets compiled for Project Phoenix, a search for extraterrestrial technological signals. In this paper we present a catalog of stellar systems that are potentially habitable to complex life forms (including intelligent life), which comprises the largest portion of the new SETI target list. The Catalog of Nearby Habitable Systems (HabCat) was created from the Hipparcos Catalogue by examining the information on distances, stellar variability, multiplicity, kinematics, and spectral classification for the 118,218 stars contained therein. We also make use of information from several other catalogs containing data for Hipparcos stars on X-ray luminosity, Ca ii H and K activity, rotation, spectral types, kinematics, metallicity, and Stromgren photometry. Combined with theoretical studies on habitable zones, evolutionary tracks, and third-body orbital stability, these data are used to remove unsuitable stars from HabCat, leaving a residue of stars that, to the best of our current knowledge, are potentially habitable hosts for complex life. While this catalog will no doubt need to be modified as we learn more about individual objects, the present analysis results in 17,129 Hipparcos 'habstars' near the Sun (75% within 140 pc), ~2200 of which are known or suspected to be members of binary or triple star systems.

(Turnbull and Tarter, 2002: 181)

(Note: there is no commentary for this activity.)

Activity

Read the introduction to the article about the search for habitable stars in Text 13. What background knowledge is provided? What definitions and facts? What is presented as the 'known' and the 'unknown' in the field? What is the hypothesis guiding the research conducted? (What do they expect or hope to find?)

Text 13

1.1. Introduction

The creation of a Catalog of Habitable Stellar Systems (HabCat) was motivated specifically by a need for an expanded target list for use in the search for extraterrestrial intelligence by Project Phoenix of the SETI Institute. Project Phoenix is a privately funded continuation of NASA's High Resolution Microwave Survey (HRMS), a mission to search for continuous and pulsed radio signals generated by extrasolar technological civilizations. HRMS consisted of an all-sky survey in the 1–10 GHz frequency range as well as a targeted search of 1000 nearby stars at higher spectral resolution and sensitivity in the 1–3 GHz range. Although Congress terminated HRMS in 1993, the SETI Institute raised private funds to continue the targeted portion of the search as Project Phoenix. Project Phoenix now carries out observations at the Arecibo Observatory in conjunction with simultaneous observations from the Lovell Telescope at the Jodrell Bank Observatory in England. The project uses a total of 3 weeks of telescope time per year and is able to observe ~200 stars per year. In the near future, the SETI Institute expects to increase the speed of its search by a factor of 100 or more. In a joint effort by the SETI Institute and the University of California at Berkeley, the Allen Telescope Array (ATA; formerly known as the

One Hectare Telescope) is currently being designed for the Hat Creek Observatory located in northern California. The ATA will consist of 350 dishes, each 6.1 m in diameter, resulting in a collecting area exceeding that of a 100 m telescope. On its current development and construction time line, the ATA should be partially operational in 2004 and fully operational in 2005. The construction of the ATA will mark an increase in telescope access and bandwidth capability sufficient to observe thousands to tens of thousands of SETI target stars per year. Hence the observing list for Project Phoenix needs to be greatly expanded from its original scope of about 2000 of the nearest and most Sun-like stars (Henry et al. 1995). The Catalog of Nearby Habitable Stellar Systems presented in this paper comprises the largest portion of SETI's new target list (to be discussed fully in a subsequent publication).

1.2. Defining habitability

Our goal is to build a catalog of stars that are potentially suitable hosts for communicating with life forms. In defining the habitability criteria for SETI target selection, we note that the development of life on Earth required (at the very least) a terrestrial planet with surfacial liquid water and certain heavy elements (e.g., phosphorus), plus an energy source (e.g., sunlight) (Alberts et al. 1994). The basic requirement of terrestrial planets suggests that there may be a lower limit on stellar metallicity for habitability (discussed in x 3.7). Given the possibility of terrestrial planets, the second requirement of liquid water means that the concept of a 'habitable zone' (HZ), i.e., that annulus around a star where the temperature permits the presence of liquid water on an Earth-like planet (investigated in detail by Kasting, Whitmire, & Reynolds 1993, hereafter K93), is a recurring theme as we evaluate the habitability of stars with different spectral types and also multiple-star systems.

An additional requirement for the development of complex life on Earth has been the continuous habitability of the planet over billions of years. Although there is evidence that simple life forms inhabited Earth as early as 0.8 billion years after Earth's formation (Schopf 1993), according to the fossil record and biomolecular clocks, multicellular life did not appear until after 3–4 billion years (e.g., Rasmussen et al. 2002; Ayala, Rzhetsky, & Ayala 1998; Wray, Levinton, & Shapiro 1996), and the emergence of a technological civilization capable of interstellar communication occurred only in the last century. The requirement for a long habitability timescale τ_{hab} strongly impacts the number of stars included in HabCat. All such stars must be older than τ_{hab}, and their HZ locations must not change by more than the HZ width over that time. However, it is not clear that the 4.6 billion year time to intelligence on Earth is a universal requirement for the appearance of interstellar communication technology (e.g., arguments made by McKay 1996).

53

Here we acknowledge that the determination of a minimum τ_{hab} for SETI targets is arbitrary, and we follow the examples of Dole (1964), Hart (1979), and Henry et al. (1995) in setting τ_{hab} 1/4 3 billion years. Combining these ideas, we define a 'habitable' stellar system as a system in which an Earth-like planet could have formed and supported liquid water throughout the last 3 billion years. For convenience, we call the host star of such a system a 'habstar'. Implicit in the definition of a habstar are concerns about metallicity, companions, stellar age and mass, and stellar variability. We expect that the habitability criteria presented below will need to be adjusted as more is learned about the formation of terrestrial planets, the origin and evolution of life on Earth, and the presence of life, if any, on other planets or moons of the solar system.

(Turnbull and Tarter, 2002: 181)

Note: The references in this passage are from the original article.

Commentary

The authors of this article review what their field agrees are the criteria for life as we know it on earth: liquid water, heavy elements, continuous life for billions of years to allow evolution of life forms into those capable of communicating with us. The authors want to establish these criteria as their guide for locating stars that might support other earth-like planets. The authors give credit to other scientists by citing their work in establishing these criteria for habitability of a planet. Also in their introduction authors explain their purpose and hypothesis, what they want to do in their research and what they expect to find. Our authors hope to build a catalogue of star systems that could be targets for those searching for signals from alien worlds. They hypothesize that even as they identify such stars, their criteria may change as knowledge in planetary science increases.

Activity

For the sake of brevity, we will not include all of the methods sections from this article. Below in Text 14 are key statements from each section describing how habitable star systems were identified. Examine the statements and identify the methods used to create their catalogue of habitable star systems.

1.3. The Hipparcos Catalogue as a starting point for HabCat

Therefore our procedure was to begin with the entire Hipparcos Catalogue and eliminate stars for which currently available data indicate nonhabitability.

2. The celestia sample

2.1.

... In order to do any assessment of the habitability of Hipparcos stars, we require some estimate of luminosity and temperature. Therefore we have included only stars for which B and V photometry and parallax were obtained, and we did not include stars whose parallax measurements were less than zero (due to large uncertainties). ...

2.2. Variability detected by Hipparcos

What about variable stars? We know that all stars are variable at some level, but how much fluctuation is tolerable to life, or to complex life? The most well-studied variation of the Sun is the 11 year sunpot cycle (also called the Schwabe cycle), and during solar minimum the Sun's total irradiance fluctuates by only ~0.02%. ... However, given that these extremely small fluctuations did have noticeable impacts on global climate, we have taken the view that stellar variability greater than 1% in luminosity would be a significant concern for habitability. ... Erring on the conservative side, we have chosen to remove all stars with detected variability: 'unsolved' variables 'microvariables' 'variability-induced movers' and stars included in the Variability Annex Part C. As for periodic variables stars classified as cataclysmic, eruptive, pulsating, rotating, or X-ray variables were eliminated, but variables identified as eclipsing binaries were retained for analysis in x 3.8. ...

2.3. Multiplicity in the Hipparcos Catalogue

... The presence of more than one star in a stellar system places limitations on where planets can form and persist in stable orbits. In order for a multiple system to be habitable to life, stable planetary orbits must coincide with the habitable zone. ... We excluded from the Celestia sample entries containing more than two resolved components. ....

Commentary (on methods section)

Scientists describe their research or experimental design in their methods section. They must show that they have controlled conditions to ensure their results are not tainted by uncontrolled factors. So they must explain how they selected their subjects, what means they used for collecting and measuring data, and how they controlled variable factors that could affect the experimental outcome.

In the article about the search for intelligent life in other solar systems, the authors describe the parameters they used to include and exclude star systems in their database. The temperature, luminosity, degree of variations in luminosity, size, and the number of sun-like stars in a solar system were all factors used to limit their search for only those star systems that met their criteria for systems that might support habitable planets.

In Text 15 the authors report the results of their first sample – the Celestia query.

Text 15

2.4. The Celestia query and resulting sample

The Celestia query resulted in a total of 64,120 stars out of the original 118,218. In Table 1, we show the exact criteria specified in the Celestia query. . . .

TABLE 1: The Celestia Query
Query Parameter Specification
Number of Stars

1. Hipparcos stars	All entries	118,218
2. Photometry	~1.037< B_V < 5.460	116,937
3. Parallax	~ > 0 mas	113,710
4. Parallax uncertainty	~/<0.3	69,301
5. Coarse variability	< 0.06 mag	4112
6. Coarse variability	0.06 to 0.6 mag	6351
7. Coarse variability	> 0.6 mag	1099
8. Variability annex	Unsolved variables (2)	5542
9. Variability annex	Light curve (not folded) (C)	827
10. Variability type (1 letter)	Microvariable (M)	1045
11. Variability type (1 letter)	Unsolved variables (U)	7784
12. Multiplicity annex	Variability induced movers (V)	288
13. Multiplicity annex	Stochastic solution (X)	1561
14. Resolved components	3 or 4	135
15. Variability type (5 letters)	E, EA, EB,EW	986
16. Combined criteria	1 AND 2 AND 3 AND 4	69,014
17. Combined criteria	NOT 14 AND (4 OR 5 OR 6)	10,576
18. Combined criteria	1 AND NOT (7 OR 8 OR 9 OR 10 OR 11 OR 12 OR 13 OR 16)	64,120

(Turnbull and Tarter, 2002: 181–2)

Commentary

The findings section presents the data found. Scientists must show whether or not what they expected, their hypothesis, is born out by the data. Data may be presented in tables and charts as well as in prose.

In the article about the search for extraterrestrial life, the authors provide the results of searching several star catalogues from which they made cuts of star systems that did not meet their criteria for systems that might support habitable planets. The resulting list included 17,129 star systems which they called the 'Habcat' – 'The Catalog of Habitable Systems' – that can serve as targets for pointing telescopes that can pick up technical signals from alien civilizations. This information is presented in three charts that provide information about the target stars, their position, size and distance from each other. The authors explain that they have used the criteria for habitable planets to trim the list of 118,218 stars in the Hipparcos Catalogue down to the 17,129 stars in the Catalog of Nearby Habitable Stellar Systems (HabCat). HabCat will serve as the list of preferred targets for targeted searches carried out by the SETI Institute from the Allen Telescope Array.

Activity

Text 16 shows the discussion section of the article about habitable star systems. Read the discussion and determine what the authors considered most important to emphasize about their findings.

Text 16

4. The catalog of habitable systems

To briefly restate our criteria for habitability, a 'habstar' must (1) be at least 3 Gyr old, (2) be nonvariable, (3) be capable of harboring terrestrial planets, and (4) support a dynamically stable habitable zone (defined by that annulus where an Earth-like planet could support liquid water on its surface). We have used those criteria to trim the list of 118,218 stars in the Hipparcos Catalogue down to the 17,129 stars in the Catalog of Nearby Habitable Stellar Systems (HabCat). HabCat will serve as the list of preferredtargets for targeted searches carried out by the SETI Institute from the Allen Telescope Array. Despite the broad array of data used to assemble this catalog, this exercise has forced us at every turn to admit that we are defining 'habitability' from a position of considerable ignorance. A complete characterization of all the stars within a few hundred (or even a few tens of) parsecs, including their masses, ages, variability, and whether they have stellar companions or planetary systems (including terrestrial planets), is simply not realizable at this time. In addition, many theoretical questions remain regarding the effects of metallicity on planet formation, the kinematics of stars and whether spiral arm crossings are truly deleterious to life forms, the effects of stellar variability (including timescales of hours, days and decades) on planet climate, the effect of stellar/giant planet companions on terrestrial planet orbital eccentricity, the effect of the stellar spectral energy distribution on the evolution of plants and other life forms, the suitability of giant planet moons for life (given expected impact rates, tidal heating, and particle radiation), etc. For SETI, this humbling situation is amplified when we consider that we have no indisputable definition for 'life' itself, to say nothing of the precise conditions that are necessary and sufficient for life to evolve into a technological civilization detectable by a SETI search program. HabCat reflects the state of our current knowledge and will evolve as we learn more about Galactic structure, the solar neighborhood, planets, life in the solar system, and the evolution of intelligence on Earth.

(Turnbull and Tarter, 2002: 181)

Commentary

The discussion section is where scientists argue their interpretation of the findings. If the investigators did not find what they expected to find, they may point to weaknesses in their research design or they may emphasize the information value of negative findings. Discussion sections also serve to emphasize the key findings that support scientists' claims and the need for more research in the area.

Our star searchers, alas, did not find intelligent life in outer space, but they did identify some good places to look. In their discussion of this project, they admit to some limitations of their methods:

Despite the broad array of data used to assemble this catalog, this exercise has forced us at every turn to admit that we are defining 'habitability' from a position of considerable ignorance. A complete characterization of all the starts within a few hundred (or even a few tens of) parsecs, including their masses, ages, variability, and whether they have stellar companions or planetary systems (including terrestrial planets), is simply not realizable at this time.

Our authors also identify key concepts that need to be defined when they admit to lacking an 'indisputable definition for "life" itself, to say nothing of the precise conditions that are necessary and sufficient for life to evolve into a technological civilization detectable by a SETI search program'.

TYPES OF STATEMENTS AND THE EVOLUTION OF SCIENTIFIC 'FACTS'

The experimental report is a complex web of linguistic and rhetorical rules that are mostly tacitly understood by writers and readers. These rules are generally intended to prevent misunderstandings that can lead audiences to make wrong assumptions. For example, one rule is that writers must not convey a certainty that isn't backed up by their evidence. Since absolute proof is usually beyond the limits of any experiment, most claims based on the evidence in one report, even when added to a growing database, cannot be proved in any absolute sense. The purpose of rules that place such limits is to ensure that those making statements do not overstate reality, that they do not imply greater certitude than their evidence actually reveals.

For example, when scientists have a hunch that the cause of a disease may be a particular virus but the evidence they report is inconclusive, they may include a statement like the following in their research report: 'While our experiments provide no conclusive evidence for a causal effect, we *suggest* that an oncovirus *be considered* as having a *possible* role in pathogenesis.'

Writers of such statements understand the rules. They know that if they are not certain or if they cannot provide substantial evidence, they must **hedge** their statements in such a way that their uncertainty is conveyed to their readers. Notice the italicized terms are those that detract from certainty. Generally, the more hedging terms contained in a statement, the more speculative it is. Those statements containing the least number of hedges are those that indicate **fact status**, that is, that the claim or condition described in the statement has been so ratified by sufficient evidence and by community agreement that it is a fact. Here is an example: 'The life cycles of flowering plants fall into three categories: annual, biennial, and perennial.'

Most often, this type of statement can be found in textbooks intended for science students. This statement was taken from *Life: The Science of Biology* (2001: 267). Textbooks convey what scientists in a given area of Science agree to be true or factual. If someone discovers an additional plant life cycle, such as what might occur in a gravity-free environment or on another planet, this textbook will be revised to reflect that change.

Discourse rules are intended to prevent misunderstandings. A speculation presented as a fact or a fact presented as a speculation will merely create confusion. Generally, the scientific literature of any scientific field is a record of that field's movement from **conjecture** to facts because that literature contains statements made according to the rules of the discourse.

Activity

Go back to the introduction of the article about star systems in Text 13. Underline or copy three statements that contain many hedging terms. Then copy three statements that do not contain hedging terms. From these six statements, what do you conclude about what the authors think is speculative and what they believe is factual about their topic?

(Note: there is no commentary for this activity.)

TRACING THE LINGUISTIC EVOLUTION OF FACTS

An excellent example of rule-governed patterns of scientific discourse can be found in the published papers on a given scientific problem. As knowledge and data increase, patterns of discourse shift accordingly. In general, a scientific field addressing one problem moves from a conjectural stage in which there may be several competing hypotheses to a gradual narrowing of possibilities to one or two explanations, and finally, to what the community agrees is textbook knowledge, which is what you are reading in your science classes.

It is possible then, to trace scientific agreement, at least the textual representation of that agreement, by analysing the changes in discourse patterns that accommodate the periods of conjecture, narrowing of hypotheses and, finally, agreement.

We can trace the evolution of facts by examining statements placed at key positions in scientific papers. For example, as we learned earlier, introductions present the knowledge base or what the community already agrees about as well as what the authors wish to add to that knowledge base. Statements describing the knowledge base in introductions are often the most certain in their tone. Statements describing experimental results also contain few hedges unless scientists are unsure about what their data reveal. Discussion sections contain more hedging words because this is where most speculation or conjectures occur; scientists are attempting to influence their readers about those matters that may not be absolutely certain, such as the implications and importance of the research.

Bruno Latour and Steve Woolgar (1987), sociologists who studied how scientists come to agree about what is factual, have given us a paradigm for examining these pattern changes in the published literature. They noted five types of statement regarding the most important claims about phenomena. These are explained in the following section.

THE CLASSIFICATION OF STATEMENT TYPES

Type 1 statements

Type 1 statements are the most speculative; they are conjectural statements marked by the use of hedges. Hedges are words such as 'might' or 'suggest'. Type 1 statements also contain **modalities**, references to human agents, to the time of discovery, or to the circumstances or conditions that are needed to make the statement plausible. These statements may also contain qualifiers such as 'high', 'low', 'novel' or any other descriptor that increases or decreases the audience's attention to some aspect of the phenomenon. The relationship expressed in Type 1 statements may follow these patterns:

> Certain conditions, i.e. data, make it highly probable that X will result.
>
> *or*
>
> The given data suggest the possibility of X.
>
> *or*
>
> Although Y, X may be possible. [Y being the conditions that could falsify the statement.]

61

Activity

Read the example in Text 17. Can you find the features of Type 1 statements?

Text 17

This statement comes from the discussion section of an early medical report on a mysterious, fatal condition among previously healthy young men in New York City:

Although the volunteers in this study may not be representative of the overall homosexual population in New York city, the strikingly high prevalence of reduced OKT4/OKT8 ratios in our subjects does suggest that this alteration may be present in a large number of homosexual men in the community.

(Kornfield *et al.*, 1983: 731)

Commentary

The writer of this article may have been much more certain of the idea of widespread infection than he is revealing here. But he followed a tacit discourse rule that urges caution in the announcement of epidemics, that disallows certainty or definitive language until all other possibilities or alternative explanations have been eliminated in public discourse. Also, he is making this statement in the discussion section where it is conventional to include more speculative statements.

Type 2 statements

Type 2 statements represent the transition from speculation to claim. That is, the writers of the paper are not just suggesting possibilities but claiming that they have the answer. These statements may begin with the words 'since' or 'if'. The basic relationship expressed in Type 2 statements is that certain conditions point to the conclusion or theory. The pattern is generally:

If condition, then X.

Text 18 is an example from a medical report on AIDS, again in the discussion section. Pick out the features that make this a Type 2 statement.

Text 18

> . . . at this time cytomegalovirus is highly suspect, in view of its prevalence among male homosexuals and its previously documented potential for immunosuppression.
>
> (Gottlieb *et al.*, 1981: 1430)

Dr Gottlieb is claiming that one 'highly suspect' causative agent in the new disease is cytomegalovirus. He does not simply state his opinion but also includes those conditions that he thinks point to the validity of the claim – that cytomegalovirus prevails among homosexuals and that it is known to suppress the immune system. He is making a claim and providing some evidence, a single voice in growing conversation about this new disease.

Type 3 statements

Type 3 statements signal an emerging agreement or a convergence of shared experience among community members. What once was speculation that moved into a claim made by one laboratory is now something that many laboratories have also reported or argued.

Still, Type 3 statements are not yet textbook facts, and they often contain reference to the conditions that make the statement true. The pattern is:

Several reports of X have appeared in the last few years.

or

Many observers have recently reported X.

63

Activity

Text 19 shows an example from a report on AIDS. Identify the features that make this a Type 3 statement.

Text 19

Unexplained chronic generalized lymphadenopathy has recently been reported in homosexual men in several metropolitan areas in the United States.

(Ewing *et al.*, 1983: 819)

Commentary

This statement is the first sentence of the introduction and offers what represents the nearest thing to a fact about AIDS that there was in 1983 when the report was published. The statement 'recently been reported' alludes to a growing consensus about the existence of an unexplained condition among homosexual men. As more reports appear, this consensus solidifies into a fact.

Type 4 statements

Type 4 statements have fact status; that is, they contain no modalities and usually represent the knowledge that eventually appears in textbooks. In these statements, phenomena are defined, and writers often use these statements to build a context for their claim. No longer is it necessary to refer to shared experience or consensus through such phrases as 'most agree' or 'several reports of'. The basic pattern for these statements is:

X is defined as.

or

X is.

The example in Text 20 comes from an AIDS report. How does the first of the two sentences fit the pattern for Type 4 statements? What do you think is the relationship between the first and second sentences?

Text 20

> Dendridic cells are close relatives of endothelial cells, by several criteria. Both can function as antigen-presenting cells. One could therefore speculate . . .
>
> (Belsito *et al.*, 1984: 1281)

Notice that the writers do not bother to spell out the 'several criteria' they allude to because they assume the readers have this knowledge. The definition in the first sentence is used to set up the following speculative statement.

Type 5 statements

Type 5 statements are the most fact-like. They are the 'taken for granted' facts that make up the implicit knowledge of a community. Often they are so taken for granted that no explicit reference to them is made. For example, research reports in microbiology do not define DNA or explain its function as the building block of life. These are the 'textbook' facts of science. They have lost the modalities of time and origin of construction and appear as if they were never proposed by anyone.

Often, these 'facts' resurface in published discourse when an anomaly has surfaced, such as young, previously healthy young men with acquired immune deficiency.

65

Activity

The statement in Text 21 contains textbook knowledge that would not normally be mentioned. Why do you think it was necessary to include this statement in a report of people with immune defects who have no known predisposition to such a problem?

Text 21

Acquired T-cell defects are well known to occur in adults with untreated Hodgkin's disease, sarcoidosis, and viral infections.

(Gottlieb *et al.*, 1981: 1425)

Commentary

This statement captures basic textbook knowledge about immunodeficiency. The statement has high fact status. Yet, as Gottlieb hopes to show, textbook knowledge about immunodeficiency does not account for what he has found in his patients who have none of these preconditions.

Evolution of statements toward 'facts'

The evolution of a discourse towards reflecting agreement among scientists goes something like this:

1 A statement about a phenomenon appears as a speculative conjecture (Type 1) near the end of an experimental report, in the discussion section.

2 Social and intellectual forces – the accumulation of reports, the corroboration of conjectural statements, an explosion of research – combine to move the statement to claim status (Type 2) which pushes it up towards the introduction of a paper.

3 As consensus builds, the statement takes on fact status and may become the first sentence of an introduction (Type 3).

4 Eventually, the conditions making the statement factually plausible drop off, and the statement becomes a definition (Type 4), appearing in the introduction or as an explanatory condition for another claim.

5   Finally, as the statement becomes incorporated into implicit know-
     ledge, it passes out of published papers altogether, existing in
     undergraduate teaching texts (Type 5).

     Latour and Woolgar (1987), who gave us this statement scheme,
traced the evolution of statements about an enzyme, TRC, from the
speculation in early laboratory investigations to the conclusive language
in subsequent research reports. It is possible to trace community agreement about other scientific phenomena by examining the published
papers in the field.

TRACING CONSENSUS ABOUT AIDS

The AIDS epidemic was officially recognized by the medical community
with the publication of three medical reports in an early December 1981
issue of *The New England Journal of Medicine*. Before that, doctors had
been treating unusual infections among homosexual men and drug
users (as well as people who belonged to neither category) that invari-
ably led to death. These infections, such as pneumocystis pneumonia,
do not affect people with normally functioning immune systems.
Though the rate of infectious disease was higher among gay men, there
was no agreement within the medical community that these infections represented a new disease. So existence of the phenomenon, a
new disease, had to be established in discourse. The next step was to
determine the cause of this disease. The evolution of agreement about
existence and cause of AIDS can be traced by examining the early
medical reports on AIDS.

Activity

In Text 22 you will find statements relating to the cause of AIDS. Identify
the statement type for each sentence. Identify all modalities and hedges
in each statement. After you have done this, write a brief response to
what you can say regarding the development of the consensus about what
causes AIDS.

Text 22

1984

The novel aspects of AIDS and the expanding insights into the epidemiology of the condition strongly support an infectious cause and suggest the involvement of an unusual agent, such as the human T-cell leukemia retrovirus.

(Scott *et al.*, 1984: 80)

1984

The preference of LAV [lympadenopathy Associated Virus] for a subset of the helper/inducer (leu-3a+ or OKT4+) T-lymphocyte population, its capacity to produce cytopathic changes in these cells, and the serum immunoglobulin profiles of antibodies to LAV in persons with AIDS or at risk from AIDS make this virus an attractive etiologic candidate.

(Laurence *et al.*, 1984: 1269)

1985

Epidemiologic data strongly suggest that the acquired immuno-deficiency syndrome (AIDS) is caused by an infectious agent, most likely a virus, which is transmitted horizontally through intimate contact.

(Schupback, 1986: 265)

1985

Recent studies indicate a close association between newly discovered human retroviruses and the acquired immunodeficiency syndrome (AIDS).

(Hirsch *et al.*, 1985: 1)

1985

Several reports suggest that human T-cell lymphotropic virus Type III (HTLV-III), the etiologic agent of AIDS, may be directly involved in these processes.

(Ho *et al.*, 1985: 1493)

1985

Human T-cell leukemia (T-lymphotropic) virus Type III (HTLV-III) is the etiologic agent of the acquired immunodeficiency syndrome (AIDS).

(Resnick *et al.*, 1985: 1498)

1988

Risk factors for infection with the human immunodeficiency virus (HIV), have been identified and widely published.

(Pizzo *et al.*, 1988: 889)

1988

Human immunodeficiency virus (HIV) infection is a progressive, lethal disease that is often accompanied by neurological disorders, including cognitive changes, progressive dementia, peripheral neuropathy, and paraplegia.

(Schmitt *et al.*, 1988: 1573)

Commentary

By closely examining the changes in discourse over time, we can trace the growing agreement about HIV infection. Early descriptions of a strange condition among gay men gradually shifted to claims about a viral infection causing a host of disorders in all human populations. What causes AIDS, at least the biologic agent, is no longer a matter of debate. What treatments work best, whether vaccines can be developed and what social and political policies can deter the spread of HIV infection continue to be explored and debated.

SUMMARY

Hopefully, you have a better understanding of the format and purpose of experimental research reports in science. If you are curious about a subject, take the plunge and try to read a report intended for scientific audiences. You may be able to determine what the authors consider 'facts' in their fields, what they believe is possible and what they are arguing based on their evidence. While much of the language may be too difficult, you may still gain a basic understanding of the main ideas.

Extension

Trace consensus and associated changes in discourse for a topic of your choice within any area of science. You might choose a disease that affects you or someone you know, a highly controversial topic, or a theoretical topic. You may log on to a specialized scientific database or you can use a more general database, but stay with scientific papers written for scientists rather than science journalism. You may limit your search to one journal in the field. Examine the introductions, discussions and conclusions for what the community considers speculative, debatable, confirmed and factual. Here are some topics if you do not have one of your own:

- embryonic stem cell research;

- dark matter (astronomy);

69

◎ intersexuality;

◎ environmental toxins;

◎ global warming.

FURTHER READING

Atkinson, Dwight, *Scientific Discourse in Sociohistorical Context: The Philosophical Transactions of the Royal Society of London, 1675–1975*, Mahwah, NJ: Lawrence Erlbaum Associates, 1999.

Bazerman, Charles, *Shaping Written Knowledge: The Genre and Activity of the Experimental Article in Science*, Madison: University of Wisconsin Press, 1988.

Carter, Ron, Angela Goddard, Danuta Reah, Keith Sanger, Maggie Bowring, *Working with Texts: A Core Book for Language Analysis*, 2nd edn, London: Routledge, 2001.

Gross, Alan, Joseph E. Harmon, Michael Reidy, *Communicating Science: The Scientific Article from the Seventeenth Century to the Present*, Oxford: Oxford University Press, 2002.

Latour, Bruno and Steve Woolgar, *Laboratory Life: The Construction of Scientific Facts*, Princeton, NJ: Princeton University Press, 1987.

SOURCES

Belsito, Donald *et al.*, 'Reduced Langerhans' Cells 1a Antigen and ATPase Activity in Patients with the Acquired Immunodeficiency Syndrome (AIDS) or AIDS-related Disorders', *New England Journal of Medicine* 310 (1984): 1279–81.

Ewing, Edwin P. *et al.*, 'Unusual Cytoplasmic Body in Lymphoid Cells of Homosexual Men with Unexplained Lymphadenopathy', *New England Journal of Medicine* 308 (1983): 819–22.

Gottlieb, Michael *et al.*, 'Pneumocystis Carinii Pneumonia and Mucosal Candidiasis in Previously Healthy Men', *New England Journal of Medicine* 305 (1981): 1425–30.

Hirsch, Martin *et al.*, 'Risk of Nosocomial Infection with Human T-cell Lymphotropic Virus III (HTLVIII)', *New England Journal of Medicine* 312 (1985): 1–4.

Ho, David D. *et al.*, 'Isolation of HTLV-III from Cerebrospinal Fluid and Neural Tissues of Patients with Neurological Syndromes Related to the Acquired Immunodeficiency Syndrome', *New England Journal of Medicine* 313 (1985): 1493–7.

Kornfield, Hardy *et al.*, 'T-Lymphocyte Subpopulations in Homosexual Men', *New England Journal of Medicine* 307 (1983): 729–31.

Laurence, Jeffrey *et al.*, 'Lymphadenopathy-associated Viral Antibodies in AIDS', *New England Journal of Medicine* 311 (1984): 1269–73.

Pizzo, Philip *et al.*, 'Effect of Continuous Intravenous Infusion of Zidovudine (AZT) in Children with Symptomatic HIV Infection', *New England Journal of Medicine* 319 (1988): 889–96.

Purves, William, Craig Heller, David Sadava, Gordon Orians, *Life: The Science of Biology*, 6th edn, New York: W.H. Freedman, 2001.

Resnick, Lionel *et al.*, 'Intra-Blood-Brain-Barrier Synthesis of HTLV-III-specific IgG in Patients with Neurological Symptoms Associated with AIDS or AIDS-related Complex', *New England Journal of Medicine* 313 (1985): 1498–1504.

Schmitt, Frederick A. *et al.*, 'Neuropsychological Outcome of Zidovudine (AZT) Treatment of Patients with AIDS and AIDS-related Complex', *New England Journal of Medicine* 319 (1988): 1573–8.

Schupback, J., 'Scientific Correspondence', *Science* 321 (1986): 119–20.

Scott, Gwendolyn B. *et al.*, 'Acquired Immunodeficiency Syndrome in Infants', *New England Journal of Medicine* 310 (1984): 76–81.

Turnbull, Margaret C. and Tarter, Jill C. 'Target Selection for SETI. I. A Catalogue of Nearby Habitable Stellar Systems', *The Astrophysical Journal Supplement Series* 145 (2002): 181–98.

Understanding
the rhetorical
in science

We have discussed in earlier units the qualities of language that make a piece of writing seem 'scientific'. These are what we might call the 'global' features of scientific language, the base lexical and linguistic elements that have developed over many years within scientific communities. When scientists employ these features in their writing, they are participating in the tradition of scientific discourse.

This tradition ensures that scientific communication is recognizable as scientific rather than poetic or political communication. Yet it does not ensure that all scientific communication is exactly the same, and it does not rescue scientists from the chore of persuading their peers. Any scientific paper is an act of rhetorical persuasion because the writers are attempting to persuade their readers that if they had collected the same data under the same circumstances, they would have come to the same conclusions.

When scientists write a paper about their research, they will not write the exact same paper another group of writers would write if they had done the same experiments and obtained the same data. This is because of the creative potential of language and the creativity of humans who use it. Just as you are faced with a daunting number of choices in your own writing – how to introduce your topic, how to organize your discussion, which evidence to employ – scientists also make highly complex choices about those and other matters when they write a paper. Here is a graphic depiction of the choices scientific writers make when writing a paper.

JOURNALS

The *journal* that will publish a scientific paper will have an important influence on how the paper is written. Some journals are more specialized and assume a high level of knowledge among the readers while other journals are more general, attracting readers from various backgrounds. The level of specialization will determine how audiences are addressed.

Activity

What differences do you notice in the two introductions shown in Texts 23 and 24? What do the passages reveal about the journals where they appeared and the differences between the two journals?

Text 23

From Bruce Balick and Adam Frank, 'The Extraordinary Death of Stars', *Scientific American* July 2004: 51–9.

Within easy sight of the astronomy building at the University of Washington sits the foundry of glassblower Dale Chihuly. Chihuly is famous for glass sculptures whose brilliant flowing forms conjure up active undersea creatures. When they are illuminated strongly in a dark room, the play of light dancing through the stiff glass forms commands them to life. Yellow jellyfish and red octopuses jet through cobalt waters. A forest of deep-sea kelp sways with the tides. A pair of iridescent pink scallops embrace each other like lovers.

For astronomers, Chihuly's works have another resonance: few other human creations so convincingly evoke the glories of celestial structures called planetary nebulae. Lit from the inside by depleted stars, fluorescently colored by glowing atoms and ions, and set against the cosmic blackness, these gaseous shapes seem to come alive. Researchers have given them such names as the Ant, the Starfish Twins and the Cat's Eye. Hubble Space Telescope observations of these objects are some of the most mesmerizing space images ever obtained.

Text 24

From Bruce Balick and Adam Frank, 'Shapes and Shaping of Planetary Nebulae', *Annual Review of Astronomy* 40 (2002): 439–86.

Perhaps the first and youngest of the many 'standard models' in astronomy to fall victim to the penetrating spatial resolution and dynamic range of the Hubble Space Telescope (HST) was that for planetary nebulae (PNe). Historically, in 1993, Frank, et al. confidently claimed that the morphologies of nearly all Pne could be understood as the evolving hydrodynamic interaction between fast winds from a central star and the nozzle formed by a dense torus of material presumably ejected earlier in the life of the central star. In 1994, the now-famous HST image of the Cat's Eye Nebula (Harrington & Borkowski 1994) mocked Frank et al.'s simple paradigm in several ways. First, no signs of dense tori were seen in close association with either member of an odd pair of orthogonal ellipsoidal features in the nebular core. Second, the HST image showed an incredible array of meticulously organized knotty or jet-like features that extant hydro models simply cannot explain with any credible set of presumed initial and boundary conditions. Insofar as our comprehension of the shapes of PNe is concerned the HSR image of NGC 6543 is redolent of the frustrating ambiguity of the Cheshire Cat.

[. . .]

PNe research is entering another renaissance. This is a time of intellectual puzzlement, debate, and play in which the imaginations of observers and theoreticians are crossing disciplinary boundaries in order to explain the morphologies and kinematics of PNe and pNEe. PNe are the testing grounds for many of these ideas thanks to their brightness, the full range of their shapes, the paucity of loal extinction, and their large numbers. The expanding constellation of interpretive ideas that has emerged to explain PNe morphologies are ready to be organized and reviewed.

Note: The references in this passage are from the original article.

Commentary

The two introductions in Texts 23 and 24 reveal a great deal about the journals where they were published. In the first introduction, the goal is to attract readers who may be unfamiliar with planetary nebula. There is no summary of specialized knowledge about nebulae, only presentation of details – what they look like and what they are named – that will be interesting to the general reader. The second introduction comes from a specialized journal read by professional astronomers. The introduction opens with one of the kinds of problems that all scientists find interesting – when a standard model is proved incorrect. Other evidence of the specialized nature of the journal includes the use of acronyms, specialized terms and citations to a body of scientific literature.

AUDIENCE

The *audiences* for any scientific paper may range from other specialists in the area to journalists writing for the popular audience. Scientists may consider only their specialized peers as their audience or they may consider wider audiences, especially if their topic is provocative or of interest to the general public. They must also consider those members of their specialized community who may be sceptical or may disagree with their findings.

Activity

What can you infer about the intended audiences for the two passages in Texts 23 and 24? What different audience needs seem to be addressed in the abstracts included above?

Commentary

Our authors, astronomers Bruce Balick and Adam Frank, have an exceptional understanding of how to communicate to various audiences. In their article for the general reader of *Scientific American*, they move from the known to the unknown by drawing from readers' familiarity with blown glass structures to introduce them to nebulae. While their article includes photos of nebulae taken by the Hubble telescope, our authors use vivid descriptive language to awaken their readers' curiosity and interest.

For their more specialized peers, our authors are more interested in stimulating interest in unsolved scientific puzzles than in the wonders of nebulae. Our authors welcome their peers to help them develop a more accurate model of planetary nebula. They present the data, proving one author's (Frank's) model wrong. Being wrong, however disappointing, is not cause for disillusionment but rather reinvigoration over a new puzzle to solve. Like the Cheshire Cat, the shape and formation of nebulae are a 'frustrating ambiguity' that our authors link to 'another renaissance', evoking a time of change and exciting new ideas. This introduction is intended to attract other scientists to research that will lead to new knowledge.

THE NATURE OF THE SUBJECT MATTER AND THE AUTHORS' PURPOSE

The *subject* and purpose of any scientific article will affect how the paper is written and structured.

Is the subject new?

Notice that for our *Scientific American* audience, planetary nebulae may be a new topic, which is why our authors introduced their beautiful structure with references to blown glass.

For the specialized audience, planetary nebulae are not new, so our authors do not worry about introducing the topic of planetary nebulae. However, what they do introduce – that what was assumed to be true about nebular structure is incorrect – is introduced as a new part of the topic to be addressed.

When scientific authors are introducing and describing a new topic or problem for scientists, they must use language to establish their subject as real and compelling.

Activity

Text 25 is an excerpt from my interview with Dr Michael Gottlieb who wrote the first report of what was eventually called AIDS. His report was published in June, 1981 in the *Morbidity, Mortality Weekly Report*. Later that year, in December, another of his reports was published in the *New England Journal of Medicine*. I interviewed Dr Gottlieb in his office in Los Angeles on 28 June 1994. What concerns did Dr Gottlieb have as he announced a new disease to the medical world?

Text 25

CR: Describe some of the issues or problems you faced in being the first to report AIDS.

DR GOTTLIEB: Well, there was the issue with my first report to *MMWR* [*Morbidity and Mortality Weekly Report*, Centers for Disease Control, USA] of the title of the report which was originally 'Pneumocystic Pneumonia in Homosexual Men'. And the CDC [Centers for Disease Control] edited it and changed it to 'Pneumocystis Pneumonia, Los Angeles'. The phenomenon was our observation of pneumocystis Carinii pneumonia in homosexual men, that was an accurate observation. All of the patients reported in the *MMWR* were openly gay, and those of us in academic medicine at that time knew very few openly gay people, and so it was an experience getting to know these first patients. And we were very accepting of their sexuality and thought that it was appropriate in the title of that report. We weren't hung up about saying homosexual in our report but the CDC was more in tune with political considerations and indeed, ultimately, AIDS proved not to be a gay disease, but our initial description really should have been reported the way we saw it. And in the *NEJM* [*New England Journal of Medicine*], the title more accurately reflected our observation.

CR: What is most important in describing a new medical problem? Do you think you were successful?

DR GOTTLIEB: I was professional. With being professional academically, you want to be convincing to people that what you're observing is something real, something new, worthy of their attention, and you write your case reports in your best medical language. You're a medical scientist and you want to report something and you want to put forward your opinions as to what its significance is, but you don't want to go beyond the data. You want to try and stick with the scientific method.

CR: Was there anything beyond the data that you felt was important but you knew you couldn't say?

DR GOTTLIEB: I don't recall whether we said it or not, but the fact that we'd seen a cluster – five cases in a relatively short time – of a new and devastating disease among men of the same sexual preference and knowing the potential for the spread of sexually transmitted disease in any highly sexually active population like gay men, there was a big worry, pretty scary. Whatever this is, it's probably going to become a major catastrophe.

CR: But you couldn't use shrill language.

DR GOTTLIEB: But we didn't have the data to say that, right. That would have been going beyond the data.

CR: Sooner or later, you were able to say those things.

DR GOTTLIEB: Sure, in CDC's surveillance, after they got our report, they turned up additional cases in New York and San Francisco, and saw that this was a national disease, not a local one. There was a gay male association and further reinforced the notion that it was extensive and serious.

Commentary

Like other physicians who were the first to treat the first patients dying of AIDS in the United States in the early 1980s, Dr Gottlieb believed that a new epidemic disease was at hand. He feared 'a major catastrophe', but with just a handful of patients, he did not have sufficient evidence to express his hunch directly. He expected his audience to come to the correct conclusion, however, since he scrupulously followed the rules of discourse, using 'the best medical language', 'without going beyond the data'.

He knew that he must persuade his readers to care about a disease killing homosexual men at a time when many doctors knew very little about homosexuality and many were biased against it. He also knew that if he did not convince physicians and medical scientists to take notice, the disease would spread. He had to establish a new medical problem in the best medical language in such a way that his physician audience would recognize their first AIDS patient and so that medical scientists would be inspired to find the cause and treatments.

Other miscellaneous concerns that may affect how a paper is written may include the following: the writers' status within the community, whether they are new to the field or well known; the publicity surrounding a topic; the relevance of the topic to the health, safety, security or environment of the general public.

Through their choices about how to address any of the above concerns, writers of scientific papers or monographs can influence how their audience thinks, not just about the data presented, but about such things as the general nature of the problem, the future of research in the area, and the authors as well as those whose work they cite. And what the majority of readers come to think, as a result of reading a paper, will influence their own work as well as the progress of their field.

RHETORIC AS PERSUASIVE ART

A scientist is like any other human being who wishes to communicate persuasively; he or she must use the art of rhetoric. We will use Aristotle's definition as a starting point:

> Rhetoric is the faculty of discovering in the particular case all the available means of persuasion.
>
> (Aristotle, 1960)

First, rhetoric, even in science, is an *art* because it requires creativity and inventiveness. It is an art because it can affect other human beings in both intended and unintended ways.

Rhetoric also involves human beings in *selecting* the words, sentence styles and formats that will work best in a given situation. A high degree of experience and expertise as well as uncontroversial subjects make this selection less conscious and troublesome for writers. But with novices or with those tackling delicate or controversial subjects or who face a large opposition, selecting the right way to communicate becomes highly conscious.

Selections are made from a range of *means* or tools that are available to a given writer for a given purpose. For example, scientists cannot select comic strips as available means of persuasion in reporting their findings. They must select words and sentence styles that fall within the realm of the scientific. In science, the available means of persuasion are not the same as those means available to the advertiser, the politician, university students asking for money from their parents. Scientists may not use overt emotional appeals; they may not make promises to their audiences that cannot be supported by their evidence; they must not exaggerate their claims or dramatize their results. They may not aggressively attack their opposition nor dismiss the work of their competitors. Yet within these confines, any scientist writing a paper still faces a daunting set of choices.

The *given case* is the rhetorical situation writers or speakers face. That situation, which includes the target audience as well as the writers' main purpose, influences how persuasive the paper will be. A paper may be credible and well written, but it may not persuade those who actually read it because the writers miscalculated their audience. A paper may contain startling insights that go unnoticed because the writing is awkward and inappropriate.

Activity

One of the most important acts of scientific rhetoric performed by a scientist in the history of modern science was Charles Darwin's publication, in several editions, of his theory of natural selection in *On the Origin of the Species*. Texts 26 and 27 show some passages from his book. What do these passages tell you about Darwin's understanding of his audience and the cultural context? What do these passages reveal about the decisions Darwin made based on the nature of his subject?

81

Text 26: A passage from Darwin's Introduction

> When on board H.M.S. Beagle, as naturalist, I was much struck with certain facts in the distribution of the inhabitants of South America, and in the geological relations of the present to the past inhabitants of that continent. These facts seemed to me to throw some light on the origin of species – that mystery of mysteries, as it has been called by one of our greatest philosophers. On my return home, it occurred to me, in 1837, that something might perhaps be made out on this question by patiently accumulating and reflecting on all sorts of facts which could possibly have any bearing on it. After five years' work I allowed myself to speculate on the subject, and drew up some short notes; these I enlarged in 1844 into a sketch of the conclusions, which then seemed to me probable: from that period to the present day I have steadily pursued the same object. I hope that I may be excused for entering on these personal details, as I give them to show that I have not been hasty in coming to a decision.
>
> (Darwin, 1859: 3)

Commentary

This very first paragraph from Darwin's great work reveals his awareness of the delicate relationship he must build between himself and his audience. The general, educated nineteenth-century reader of Darwin's work would have been both curious about and sceptical of his claim that living creatures have evolved gradually over millions of years to their current state. The question of human evolution would have caused great discomfort for many people who had been raised on the creation story. The person brave enough to argue against centuries of creationist teaching could easily be dismissed as a heretic and a kook, certainly one not to be trusted. Darwin seems quite aware of his audience's likely dispensation towards him. To quell their distaste and scepticism, Darwin begins immediately to create a sympathetic impression of himself. He takes his readers with him on board the HMS *Beagle* and shares his reactions to what he has seen, his singular curiosity, his patient study. He emphasizes the patient and cautious approach he has taken in order to show his reader that he can be trusted because he is not rash and impetuous.

Text 27: A passage from Darwin's Conclusion

I see no good reasons why the views given in this volume should shock the religious feelings of any one. It is satisfactory, as showing how transient such impressions are, to remember that the greatest discovery ever made by man, namely, the law of the attraction of gravity, was also attacked by Leibnitz, 'as subversive of natural, and inferentially of revealed, religion'. A celebrated author and divine has written to me that 'he has gradually learned to see that it is just as noble a conception of the Deity to believe that He created a few original forms capable of self-development into other and needful forms, as to believe that He required a fresh act of creation to supply the voids caused by the action of His laws'.

(Darwin, 1859: 120)

Commentary

Here Darwin attempts to reconcile his religious readers' views with his own. By referring to the law of gravity as having once been attacked as subversive, Darwin hopes his readers will conclude that the theory of natural selection is, like Newton's theory, both reasonable and non-threatening. Using the testimonial of 'a celebrated author and divine' who has found reason to accept natural selection allows Darwin to make his own argument but through the quotation of another.

ANALYSING SCIENTIFIC ARGUMENTS

The arguments that are made in scientific communication must always be based on evidence or data in some form. But sometimes, the relationship between the data or evidence from laboratory research and the conclusion or claims scientists make are not obvious to all audiences. Scientific reasoning draws on patterns of reasoning that we can find in all areas of human communication. The following are some general patterns of reasoning and argumentation. Examples provided come from both general and scientific communication.

1   Reasoning from definition: arguments can be based on a definition of a thing or a process that either the audience will accept as true or that the writer hopes will be accepted as true.

Here is an example from our Tarter and Turnbull article about the search for habitable star systems. The authors base their claim of having found a number of possibly habitable star systems on this definition:

> We define a 'habitable' stellar system as a system in which an Earth-like planet could have formed and supported liquid water throughout the last 3 billion years.

2   Reasoning from relationships. Arguments can be based upon the relationships (comparisons, analogies and metaphors) between things.

Here is an example from Balick and Frank's article about planetary nebulae. They compare the ambiguous nature of nebular structure to the Cheshire Cat in *Alice in Wonderland*:

> Insofar as our comprehension of the shapes of Phe is concerned, the HSR image of NGC 6543 is redolent of the frustrating ambiguity of the Cheshire Cat.

3   Reasoning from circumstances. Arguments can be based on circumstances or situations that seem to indicate something. These arguments can be aimed at the future.

In the Tarter and Turnbull article, the authors point to the 'circumstance' of size of a planet as an indication of its ability to support life:

Of the 17,163 stars remaining in HabCat, there are 55 stars known to host 65 total planets. With the smallest minimum planet mass of 0.12 Jupiter masses (HD 49674), all of these planets are likely to be gas giants and are thus unlikely to support Earth-like life.

The authors point to additional 'circumstances' that would indicate habitability:

However, these planetary systems may still be habitable if (a) the giant planet does not interfere with the dynamic stability of the HZ or (b) the giant planet occupies the HZ throughout its orbit, giving rise to potentially habitable moons.

4   Arguments from testimony. Arguments can be based on testimony from professional experts who ought to have some credibility with the topic. Testimony can also come from people who have no real credibility with the topic but who are respected nonetheless.

Here are Tarter and Turnbull employing 'testimony' from other scientists to suggest the potential habitability of moons:

The potential habitability of moons is questionable, considering the high-radiation environment of a giant planet, possible gravitational focusing of large impactors by the giant planet, and the large eccentricities of most known extrasolar giant planets. Williams, Kasting, & Wade (1997) have shown that the effects of radiation could be avoided if the moon has an Earth-like magnetic field, and work by Williams & Pollard (2002) suggests that planets in eccentric orbits will still be habitable, as long as the stellar flux averaged over 1 year at the moon's surface is comparable to that of a circular orbit.

Activity

Read the passages shown in Texts 28 to 31 and identify Darwin's use of arguments from definition, relationship, circumstance and testimony.

85

Text 28

> Changed habits produce an inherited effect as in the period of the flowering of plants when transported from one climate to another. With animals the increased use or disuse of parts has had a more marked influence; thus I find in the domestic duck that the bones of the wing weigh less and the bones of the leg more, in proportion to the whole skeleton, than do the same bones in the wild duck; and this change may be safely attributed to the domestic duck flying much less, and walking more, than its wild parents.
>
> (Darwin, 1859: 7)

Commentary

In this example, Darwin wants to support his claim that traits that develop as a result of adaptation to an environment are inherited by subsequent offspring. He uses the *argument from circumstance* in this passage. The circumstance is his observations that the legs and wings of domestic and wild ducks differ in weight. This difference, he argues, *indicates* that different environments or habits produced the leg and wing weights which were then inherited by subsequent generations.

Text 29

> From looking at species as only strongly marked and well-defined varieties, I was led to anticipate that the species of the larger genera in each country would oftener present varieties, than the species of the smaller genera; for wherever many closely related species (i.e., species of the same genus) have been formed, many varieties or incipient species ought, as a general rule, to be now forming. Where many large trees grow, we expect to find saplings. Where many species of a genus have been formed through variation, circumstances have been favourable for variation; and hence we might expect that the circumstances would generally still be favourable to variation. On the other hand, if we look at each species as a special act of creation, there is no apparent reason why more varieties should occur in a group having many species, than in one having few.
>
> (Darwin, 1859: 26–7)

Commentary

Here Darwin is actually *arguing from a definition* of species in order to make sense of variety. He insists that the old definition of species, 'as a special act of creation' does not explain why individuals within a species vary. Darwin exhibits the use of definition as a thinking device or theory as well. He says that by 'looking at species as only strongly marked and well-defined varieties, I was led to anticipate', that more variety would result from larger genera than smaller. So this example illustrates definition as a mode of thought as well as rhetoric.

Text 30

> As man can produce, and certainly has produced, a great result by his methodical and unconscious means of selection, what may not natural selection effect? Man can act only on external and visible characters: Nature, if I may be allowed to personify the natural preservation or survival of the fittest, cares nothing for appearances, except in so far as they are useful to any being. She can act on every internal organ, on every shade of constitutional difference, on the whole machinery of life. Man selects only for his own good; Nature only for that of the being which she tends.
>
> (Darwin, 1859: 38–9)

Commentary

In this example, Darwin uses *argument from relationship* in order to persuade his readers of the internal consistency and reasonableness of natural selection. Throughout the book, he uses a different kind of argument from relationship we see here regarding the similarity of animal breeding and natural selection. In other parts of the book, Darwin uses animal breeding by humans as an analogy to natural selection in order to demonstrate the reasonableness of his ideas. But here his argument from relationship is about contrast and distinctions between natural selection and animal breeding. Natural selection is distinguished from animal breeding by how it favours the organism. Man's breeding of animals is only based on 'external and visible' characteristics whose intentional genetic manipulation may not benefit the species at all.

87

Text 31

Analogy would lead me one step further, namely, to the belief that all animals and plants are descended from some one prototype. But analogy may be a deceitful guide. Nevertheless all living things have much in common, in their chemical composition, their cellular structure, their laws of growth, and their liability to injurious influences. We see this even in so trifling a fact as that the same poison often similarly affects plants and animals; or that the poison secreted by the gall-fly produces monstrous growths on the wild rose or oak-tree. With all organic beings, excepting perhaps some of the very lowest, sexual reproduction seems to be essentially similar. With all, as far as is at present known, the germinal vesicle is the same; so that all organisms start from a common origin. If we look even to the two main divisions – namely, to the animal and vegetable kingdoms – certain low forms are so far intermediate in character that naturalists have disputed to which kingdom they should be referred. As Professor Asa Gray has remarked, 'the spores and other reproductive bodies of many of the lower algae may claim to have first a characteristically animal, and then an unequivocally vegetable existence'. Therefore, on the principle of natural selection with divergence of character, it does not seem incredible that, from some such low and intermediate form, both animals and plants may have been developed; and, if we admit this, we must likewise admit that all the organic beings which have ever lived on this earth may be descended from some one primordial form. But this inference is chiefly grounded on analogy, and it is immaterial whether or not it be accepted.

(Darwin, 1859: 212)

Commentary

In this example, Darwin makes explicit *use of analogy* in order to argue that all living creatures 'descended from some one primordial form'. Even though he admits that 'analogy can be a deceitful guide', Darwin goes on to apply analogy, the likenesses among very different genera of living creatures, to argue for the likelihood that all genera evolved from a single form. He also employs *argument from testimony* when he includes a quotation from a well-known and respected professor, Asa Gray.

RHETORIC AND AIDS

An important rhetorical situation in science centres on the race to find the cause of AIDS. As AIDS cases were beginning to emerge around the world in 1983, biomedical scientists were searching furiously for the cause. Based on several lines of suggestive evidence, a few medical scientists believed in 1983 that the cause of AIDS was viral. A well-known and highly respected American scientist, Robert Gallo at the National Institutes of Health, suspected that the cause of the new disease was a **retrovirus**. Gallo also believed that the cause of AIDS was a special type of retrovirus that he had discovered in the late 1970s, Human T-cell Leukemia Virus (HTLV) that causes a type of leukemia in different parts of the world.

Dr Gallo is the acclaimed, 'Father of Modern Retrovirology', the discoverer of the first human retrovirus and during the 1980s the director of the prestigious Laboratory of Tumor Cell Biology at the National Cancer Institute. Because of Gallo's reputation and because he had already published reports in which he argued that his retrovirus was the cause of AIDS, there were many scientists who believed Gallo was correct.

However, scientists at the Pasteur Institute in Paris in the laboratory of Jean Luc Montagnier believed Gallo was wrong. They had isolated a retrovirus from the blood of AIDS patients that they felt was unlike the HTLVs.

By 1984, both Gallo and Montagnier were ready to publish papers in *Science*. What available means of persuasion did each team of scientists select in order to persuade the world that they had discovered the AIDS virus?

This was an enormously serious rhetorical situation for several reasons. The most important was that the identification of the true cause of AIDS would lead to a blood test allowing early diagnosis and helping to prevent new infections. Blood banking companies could also test blood donations in order to protect the blood supply. To those who first discovered the cause would go not only the applause but the patent and billions of dollars. Both the United States National Institutes of Health and the Pasteur Institute stood to gain significant revenue. The scientist who could be called the discoverer of the cause of AIDS might even be worthy of a Nobel Prize. The motivations behind AIDS research were humanitarian and scientific as well as economic and political.

Activity

Text 32 shows some quotations from various members of the American and the Pasteur research teams. What do these quotations indicate about the rivalry between the two teams that were striving to identify the cause of AIDS? What do they reveal about each teams' views about scientific rhetoric?

Text 32

I have learned more of politics than of science during all this. I never thought I would have to be a good salesman in order to be heard.

(Jean Luc Montagnier, co-discoverer of HIV in Shilts, 1987: 496)

We were convinced of our data in 1983 but we were also, at that time, a little bit shy. A little differential. We hoped our readers would put two and two together. I think we don't really have a reason to be aggressive, between scientists. We just have to make science.

(Francois Barre-Sinoussi, lead author of the first French paper on virus isolation from AIDS patients in interview with the author, Washington, DC, 23 July 1995)

Montagnier did not conclude that their virus was the cause of AIDS. His style of presentation was matter of fact.

(Gallo, 1991)

Look at his papers in 1983 and 1984. It doesn't claim that he had the AIDS virus. I'm the first person who said this is the cause of AIDS, I'm positive, and I have the blood test to prove it.

He [Montagnier] was playing at indirection, you know, not making any conclusions. Well, you can make no conclusions for two reasons. Either you don't have the thought to properly make the logic or you don't have the data. I mean if the conclusion is there, you're stupid if you don't make it.

(Robert Gallo in interview with the author, National Cancer Institute, 3 August 1995)

One of the several things I've learned from this experience is how to write papers for Americans. In general, papers coming from American labs are more aggressive than the way we write papers in France.

(Francoise Barre-Sinoussi in interview with the author, Washington, DC, 23 July 1995)

The French and American laboratory teams held very different assumptions about scientific rhetoric. The team headed by Jean Luc Montagnier clearly assumed that their readers would draw the correct conclusions about their evidence without any overt rhetorical persuasion. Robert Gallo, in contrast, understood that aggressive promotion of an idea is necessary for readers to grasp its significance.

RHETORIC AND SCIENTIFIC NETWORKS

Rhetorical processes in science serve as important sources of influence. The choices a team of scientists makes when they write research reports for their colleagues in other laboratories will become very important to several social processes in science.

Sociologist Bruno Latour (1987) views science as a social process relying on politics and rhetoric to create alliances and networks of influence. He helps us understand these social processes in science that are shaped and maintained by rhetoric.

1	First, according to Latour, rhetoric used by scientists helps create networks of alliances among scientists. These networks provide support and collaboration for the scientists that belong to it. By showing that they are part of a network or by attracting new allies to the network, scientists assert their alliances and allegiances.

Text 33 shows the introduction to the article reporting a list of habitable star systems we saw in Unit two. What networks or collaborative efforts do the authors describe?

Text 33

From 'Target Selection for SETI. I. A Catalog of Nearby Habitable Stellar Systems' by Margaret C. Turnbull and Jill C. Tarter.

The creation of a Catalog of Habitable Stellar Systems (HabCat) was motivated specifically by a need for an expanded target list for use in the search for extraterrestrial intelligence by Project Phoenix of the SETI Institute. Project Phoenix is a privately funded continuation of NASA's High Resolution Microwave Survey (HRMS), a mission to search for continuous and pulsed radio signals generated by extrasolar technological civilizations. HRMS consisted of an all-sky survey in the 1–10 GHz frequency range as well as a targeted search of 1000 nearby stars at higher spectral resolution and sensitivity in the 1–3 GHz range. Although Congress terminated HRMS in 1993, the SETI Institute raised private funds to continue the targeted portion of the search as Project Phoenix. Project Phoenix now carries out observations at the Arecibo Observatory in conjunction with simultaneous observations from the Lovell Telescope at the Jodrell Bank Observatory in England. The project uses a total of 3 weeks of telescope time per year and is able to observe ~200 stars per year. In the near future, the SETI Institute expects to increase the speed of its search by a factor of 100 or more. In a joint effort by the SETI Institute and the University of California at Berkeley, the Allen Telescope Array (ATA; formerly known as the One Hectare Telescope) is currently being designed for the Hat Creek Observatory located in northern California. The ATA will consist of 350 dishes, each 6.1 m in diameter, resulting in a collecting area exceeding that of a 100 m telescope. On its current development and construction time line, the ATA should be partially operational in 2004 and fully operational in 2005. The construction of the ATA will mark an increase in telescope access and bandwidth capability sufficient to observe thousands to tens of thousands of SETI target stars per year. Hence the observing list for Project Phoenix needs to be greatly expanded from its original scope of about 2000 of the nearest and most Sun-like stars (Henry et al. 1995). The Catalog of Nearby Habitable Stellar Systems presented in this paper comprises the largest portion of SETI's new target list (to be discussed fully in a subsequent publication).

Commentary

Searching for intelligent life in the universe requires collaboration, a network of support rather than the singular efforts of a few observers. While all earth-lings would be extremely interested to know if we have company in the universe, governments are reluctant to fund the search. There are other more pressing matters to attend to. Thus, 'Project Phoenix' is privately funded and necessitates the collaboration of a network of observatories. The authors of this article hope to build more support for their mission, to attract more astronomers to join the effort. Thus, they optimistically describe future opportunities – expectations of increased search speed, increased telescope capacity – to attract fellow scientists to their work.

2 | Rhetoric also helps networks of scientists work toward estab-lishing some solid ground of agreements or ideas that everyone takes for granted.

Activity

Here, again, in Text 34 is our article about searching for life in the universe. Notice how the authors try to establish an agreement about the age of habitable star systems.

Text 34

Although there is evidence that simple life forms inhabited Earth as early as 0.8 billion years after Earth's formation (Schopf 1993), according to the fossil record and biomolecular clocks, multicellular life did not appear until after 3–4 billion years (e.g., Rasmussen et al. 2002; Ayala, Rzhetsky, & Ayala 1998; Wray, Levinton, & Shapiro 1996), and the emergence of a technological civilization capable of interstellar communication occurred only in the last century. The requirement for a long habitability timescale τ_{hab} strongly impacts the number of stars included in HabCat. All such stars must be older than τ_{hab}, and their HZ locations must not change by more than the HZ width over that time.

Commentary

Notice how the authors are attempting to establish one solid agreement – the age of the earth at the time multicellular life appeared. They cite a number of previously published articles whose authors also agree with the 3–4 billion year time period. Establishing an agreement about the age of stars to observe is very important because such an agreement about age will limit the number of stars in the database and focus the community's attention on the stars most likely to support habitable planets.

3	Rhetoric enters into the ways scientists position themselves with respect to potential allies in the network.

As Latour points out, this is an entrepreneurial game. Communication is central to enlisting allies, largely through such elements as citations and references to the literature of the field. The authors of the star article are careful to refer to published work in evolutionary biology, archaeology, as well as astronomy in order to establish basic agreements as well as to create their network of alliances that may support their work.

4	Once the community has reached agreement about what it considers 'facts', life seems harmonious until a few in the field begin to question or oppose what everyone assumes to be true.

In the quest of intelligent life in the universe, agreements about what makes a star system habitable come from what scientists have learned about how life emerged and developed on earth. Even though the authors of the article about the habitable star database have applied knowledge of the evolution of life on earth to identify star systems that might support life, they also question the assumption that life on earth is the same as life on other planets. In their introduction, they admit their 'ignorance'.

Despite the broad array of data used to assemble this catalog, this exercise has forced us at every turn to admit that we are defining 'habitability' from a position of considerable ignorance. . . . For SETI, this humbling situation is amplified when we consider that we have no indisputable definition for 'life' itself, to say nothing of the precise conditions that are necessary and sufficient for life to evolve into a technological civilization detectable by a SETI search program.

Our authors want to remind their network of peers that the very definitions used to define life may be flawed and that new ideas and new evidence are needed for the mission to find intelligent life in the universe.

5 | These communicative processes, according to Latour, create webs of agreements and relationships so strong that certain ideas, objects, facts become **black-boxed**, and are thereafter no longer seen as speculation or debatable issues.

There are plenty of examples in all areas of science. What is contained in textbooks is usually what is considered 'black-boxed', no longer a matter of debate. What leads up to 'textbook' fact is a dynamic social process in which members of a scientific community gradually, through both linguistic practices and shared experience, settle upon consensus.

What counts as 'black-boxed' issues in our article about habitable star systems? Probably, the most important black-boxed issue is not even stated: the idea that we humans ought to be devoting energy and resources to searching for intelligent life in the universe. This is a taken-for-granted view that our authors assume no longer needs to be asserted to readers of the article.

6 | Sometimes 'black-boxed' facts or issues re-emerge as debatable issues when cultural or empirical factors press them into the arena once again.

Obviously, not everyone agrees that the search for intelligent life in the universe is worth the trouble and the money. According to our article, the United States Congress ended its funding of the Phoenix Project which now depends upon private funding. But even the scientific community may come to question the value of the project if other more pressing problems arise that deserve more attention than the Phoenix Project. In the event that considerable resistance to the search for habitable star systems arises among scientists, our authors, Jill Tarter and Margaret Turnbull, will have to employ good arguments to persuade their peers to continue the search.

SUMMARY

To say that scientific communication is rhetorical and persuasive is not to say that scientists behave like politicians and advertisers. The rhetorical persuasion that convinces us to buy a brand of automobile or to vote for a particular candidate often appeals to our deepest needs and desires to move us to act. Science relies on reasonable rather than emotional rhetoric, on carefully constructed arguments that support conclusions drawn from the available evidence. Rhetoric is necessary in science since very few scientific claims are self-evident. Most claims are not self-evident; they must be supported with arguments that connect the data, whatever they are, with the interpretation and conclusion being proposed. One cannot simply provide a data chart in place of a fully developed scientific paper, with introduction, methods, results and discussion, because few readers would see the importance of the data or they would come to different conclusions. A good scientist is not only a good technician, a good experimenter, but also a good communicator.

Extension

1   Explore Charles Darwin's *On the Origin of Species* in more depth than we have here. Read his chapter on natural selection and identify his main arguments and his appeals to his audience.

2   Find experimental reports about black holes from different decades. Read the papers to discover if any agreements solidify into 'black-boxed' issues. Do they re-emerge as matters of debate?

FURTHER READING

Aristotle, *The Rhetoric of Aristotle*, trans. Lane Cooper, Englewood Cliffs, NJ: Prentice Hall, 1960.

Darwin, Charles, *On the Origin of Species*, London: John Murray, 1859.

Galo, Robert, *Virus Hunting: AIDS, Cancer, and the Human Retrovirus. A Story of Scientific Discovery*, New York: Basic Books, 1991.

Gross, Alan, *The Rhetoric of Science*, Cambridge, MA and London: Harvard University Press, 1990.

Latour, Bruno, *Science in Action: How to Follow Scientists and Engineers through Society*, Cambridge, MA: Harvard University Press, 1987.

Prelli, Lawrence, *A Rhetoric of Science: Inventing Scientific Discourse*, Columbia: University of South Carolina Press, 1989.

Reeves, Carol, 'Rhetoric and the AIDS Virus Hunt', *Quarterly Journal of Speech* 84.1 (February 1998): 1–22.

Shilts, Randy, *And the Band Played On: Politics, People, and the AIDS Epidemic*, New York: St Martin's Press, 1987.

Science and culture: the interaction of discourses

So far, we have learned how the language of science works within science. Now it is time to explore how the language and discourse of science interact with and influence communities outside science and how the culture of those outside communities influences science.

Any discourse is both a result of and a cause of culture. The specialized discourses of scientific fields both result from and contribute to the collected values and operations that make up scientific culture. Likewise, how other human communities – our religious, social, political and personal communities – come to talk about and define certain topics, results from and contributes to the values of those communities. In many ways, scientific discourses and other societal discourses shape each other. Sometimes, scientific discourse shapes the way we think and talk about a topic, but other times, wider cultural or social norms shape the way scientists think and talk about a topic.

At times, religious, political and scientific communities have a stake in significant problems. Each of these communities employs its institiutional discourse, its way of describing and categorizing a problem. When problems touch all of these communities – such as AIDS and global warming – these various discourses collide and the people using them must either accept or confront the cultural norms conveyed discursively.

The diagram below illustrates these relationships:

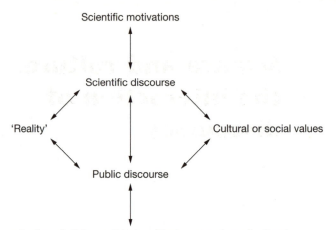

The terms in this diagram are defined as follows:

–   *Reality* is whatever object or process that is being described, explained or debated in discourse.

–   *Scientific motivations* are those interests particular to science, as a discipline, an institution and a culture. Disciplinary motivations include the aim to develop technology and methods that will broaden research scope and lead to new discoveries. Institutional motivations include the interest in protecting and maintaining the infrastructure of science that involves science education, funding, leadership and accreditation. Finally, science is a culture in which its members adhere to particular beliefs and standards of behaviour and employ a language coloured by those beliefs and standards.

–   *Scientific discourse* is the rule-governed development of descriptions and definitions of whatever 'reality' is being addressed by scientists.

–   *Public or social discourse* is the developing or developed social or public description and definition of a 'reality' that is being examined by science but which is of general public interest.

–   *Public motivations* are those interests that affect the general good and participate in the overall formation of communities.

–   *Cultural or social values* are those beliefs and practices that underlie public institutional motivations and that establish common interests and identities among members of a community.

When both science and the public address a 'reality', whether it be a disease, a space shuttle disaster, biological weapons or gene therapy, scientists' views of and the way they talk about this 'reality', may be shaped by scientific as well as public motivations and cultural values. Sometimes, there are no existing social beliefs attached to a 'reality' because it exists only as a scientific rather than a public problem. Sometimes a subject is so fraught with social prejudices and associations that social discourse has more power than science in shaping the meanings of the subject.

We will examine these relationships individually.

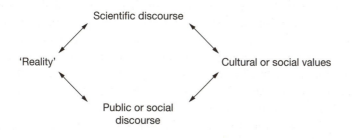

The way we come to talk about reality, whether we are scientists or not, is shaped by our experience with that reality as well as by the language we use to describe it. Our growing experience with 'reality' may be offset by the constraints in our language that disallow new views to develop. A common nineteenth-century view of the 'reality' of womanhood existing in many Western societies and still evident in some societies was that women's role was purely domestic. Despite ongoing 'experience' of the 'reality' of women's intelligence, strength, creativity and determination, the prejudices embedded in our cultural traditions and our discourse disallowed a vision that permitted women to be educated and trained in careers.

Activity

Text 35 shows the introduction to an essay by a scientist for *Nature* in 1880. Why is this passage a good example of the interaction between 'reality' (in this case women's educability), scientific discourse (testing, experimentation) and cultural values?

Text 35

From S. Tolver Preston, 'Evolution and Female Education', *Nature* 23 September 1880: 485–6.

Recently, we have heard the argument that women and common workers ought to be educated, not just in their domestic or industrial responsibilities but in the liberal arts and sciences. The argument for women's education relies on the assumption that women can learn the complicated algorithms of mathematics and the tedious conjugations of Latin. The other assumption is that they would put that knowledge to use in their household dominion. The argument for the higher education of the common worker also assumes the capabilities and usefulness of their mentalities. In both cases, these assumptions ought to be first examined with cold objectivity. We know the female has talents that males do not have in the concern of domestic life, the raising of children, the gentle care and design of home life. And we admire the physical strength and clever efficiency of honest workers. But what do we know of their mental agility? Should we not first investigate and test their potential for learning through careful experimentation before we put them to the task and risk the shame of failure?

(Note: there is no commentary for this activity.)

This diagram illustrates the interaction between scientific and public discourse that you might have noticed in Text 35:

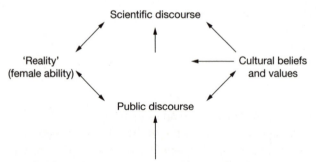

When objects or processes in nature are introduced to us by science, scientific discourse, which may be entirely appropriate within science, may shape a public discourse that is not at all appropriate. Scientists may respond by attempting to redirect the way the public talks about the issue. A good example is the way the name of the epidemic disease AIDS and those who are ill have been consciously changed over the years by scientists and patient groups responding to the damaging conceptions in public discourse.

DISCURSIVE HEGEMONIES

Hegemony occurs when one discourse – scientific or public discourse (religious, moral, economic, political) – holds sway, for a finite period, over the way we come to talk about and understand a given reality.

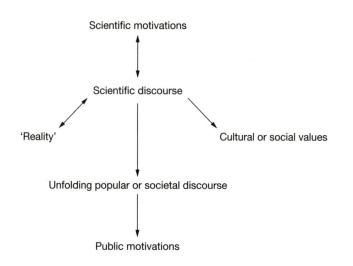

This diagram illustrates *the sway of scientific discourse over popular discourse*. The scientific discourse, developing out of growing experience with 'reality' combined with the pressures from scientific motivations as well as wider cultural values, leads to the development of popular discourse and informs the conceptions of 'reality' in public discourse.

Why does this hegemony occur? Often it is because experience with this reality begins among scientists while the rest of us have no experience and thus have developed no cultural or social values surrounding it.

When scientists introduce the public to new diseases, new astrophysical discoveries, new technological breakthroughs, we rely, at least at first, on the early conceptions that scientists provide in their language describing the new. That is why scientists must be very careful about how they introduce new phenomena. As with the early conceptions the public had about AIDS in the beginning of the epidemic, scientists' descriptions can cause wrong assumptions and destructive attitudes.

On some issues, scientific discourse is a master discourse that carries a ring of authority in debates about social policy regarding issues that have been dominated by science. When scientific discourse enters social debates as a master discourse, it can be recognized by the following elements or appeals:

◎ Scientific methods (careful investigation, experimentation, testing, analysis) are viewed as always objective, valid and beneficial.

◎ Scientific findings are viewed as leading to inevitably positive results.

◎ Scientists and science are ethical.

James D. Watson, along with Francis Crick, published the first model of DNA. Their contribution led to the field of microbiology, discoveries about the genetic causes of disease, the mapping of the human genome, and other advances. Watson has been a long time advocate of genetic engineering to eliminate pain and suffering as well as to enhance the abilities of human beings. In his speeches, Watson exploits his status as a well-known scientist and his audience's respect for scientific culture to support his argument for germ-line genetic engineering, genetic change that would be inherited in subsequent generations.

Activity

Text 36 shows an extract from an essay by Watson written for a popular audience. What aspects of scientific discourse do you find in this essay that assist him in making a case for genetic enhancement of human beings?

From James D. Watson, 'All for the Good', *Academic Search Elite* 153.1 (1 November 1999).

Why genetic engineering must soldier on

Though most forms of DNA manipulation are now effectively unregulated, one important potential goal remains blocked. Experiments aimed at learning how to insert functional genetic material into human germ cells – sperm and eggs – remain off limits to most of the world's scientists. No governmental body wants to take responsibility for initiating steps that might help redirect the course of future human evolution. These decisions reflect widespread concerns that we, as humans, may not have the wisdom to modify the most precious of all human treasures – our chromosomal 'instruction books'. Dare we be entrusted with improving upon the results of the several million years of Darwinian natural selection? Are human germ cells Rubicons that geneticists may never cross?

Unlike many of my peers, I'm reluctant to accept such reasoning, again using the argument that you should never put off doing something useful for fear of evil that may never arrive. The first germ-line gene manipulations are unlikely to be attempted for frivolous reasons. Nor does the state of today's science provide the knowledge that would be needed to generate 'superpersons' whose far-ranging talents would make those who are genetically unmodified feel redundant and unwanted. Such creations will remain denizens of science fiction, not the real world, far into the future. When they are finally attempted, germ-line genetic manipulations will probably be done to change a death sentence into a life verdict – by creating children who are resistant to a deadly virus, for example, much the way we can already protect plants from viruses by inserting antiviral DNA segments into their genomes.

If appropriate go-ahead signals come, the first resulting gene-bettered children will in no sense threaten human civilization. They will be seen as special only by those in their immediate circles, and are likely to pass as unnoticed in later life as the now grownup 'test-tube baby' Louise Brown does today. If they grow up healthily gene-bettered, more such children will follow, and they and those whose lives are enriched by their existence will rejoice that science has again improved human life. If, however, the added genetic material fails to work, better procedures must be developed before more couples commit their psyches toward such inherently unsettling pathways to producing healthy children.

Moving forward will not be for the faint of heart. But if the next century witnesses failure, let it be because our science is not yet up to the job, not because we don't have the courage to make less random the sometimes most unfair courses of human evolution.

Commentary

Watson clearly believes that science, rather than religion or government, ought to lead in making public policy about genetic engineering. In his essay, scientific authority reigns. There is much more good news than bad; we do not see doom and gloom but a bright future with less pain and suffering thanks to courageous scientists. Watson glosses over the very real possibility that the first efforts at germ-line genetic manipulation will create gross abnormalities. Watson has great faith in the ethics of scientists who would begin germ-line genetic alterations for justifiable rather than 'frivolous' reasons. He assumes that scientists are the most capable of distinguishing the justifiable from the 'frivolous'. His assumptions grow out of scientific culture, embedded in public scientific discourse. We should always be wary of this discourse and maintain a healthy scepticism of anyone who wants us to support enormous enterprises based on the assumption that science always 'improves life' and that scientists are always ethical.

SOCIETAL DISCOURSES MAY HOLD
SWAY OVER SCIENCE

Sometimes popular or social discourses about some topics may have more influence on us than does science and its discourses. When subjects are deeply linked to moral or religious beliefs, political motivations or economic necessities, science may not have as much influence. In some cases, science may be constrained by society and, thus, limited in its ability to address certain problems.

This diagram illustrates societal hegemony over science and its discourse.

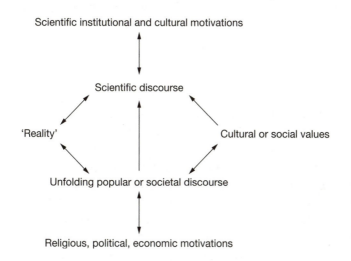

Scientific institutional and cultural motivations

Scientific discourse

'Reality'   Cultural or social values

Unfolding popular or societal discourse

Religious, political, economic motivations

Religion and science

A famous example of religion's hegemony over science occurred when the Catholic church was at loggerheads with astronomers such as Copernicus and Galileo over the nature of our solar system.

Activity

Text 37 shows excerpts of Galileo's Letter to the Church (1632) in which he defended Copernican theory that the earth and the other planets orbit the sun. How does Galileo attempt to convince the clergy? How does he try to avoid minimizing biblical authority by employing religious discourse?

107

Text 37

Translated from Galileo's Letter to the Church, 1632.

. . . I think in the first place that it is very pious to say and prudent to affirm that the Holy Bible can never speak untruth – whenever its true meaning is understood. But I believe that nobody will deny that it is often very abstruse, and may say things which are quite different from what its bare words signify. Hence in expounding the Bible if one were always to confine oneself to the unadorned grammatical meaning, one might fall into error.

[. . .]

. . . I think that in discussions of physical problems we ought to begin not from the authority of scriptural passages, but from sense-experiences and necessary demonstrations; for the Holy Bible and the phenomena of nature proceed alike from the divine Word, the former as the dictate of the Holy Ghost and the latter as the observant executrix of God's commands. It is necessary for the Bible, in order to be accommodated to the understanding of every man, to speak many things which appear to differ from the absolute truth so far as the meaning of the words is concerned. But Nature, on the other hand, is inexorable and immutable; she never transgresses the laws imposed upon her, or cares a whit whether her abstruse reasons and methods of operation are understandable to men. For that reason it appears that nothing physical which sense-experience sets before our eyes, or which necessary demonstrations prove to us, ought to be called in question (much less condemned) upon the testimony of biblical passages which may have some different meaning beneath their words. For the Bible is not chained in every expression to conditions as strict as those which govern all physical effects; nor is God any less excellently revealed in Nature's actions than in sacred statements of the Bible.

[. . .]

... I do not feel obliged to believe that that same God
who has endowed us with senses, reason: and intellect has
intended to forgo their use and by some other means to give
us a knowledge which we can attain by them. He would not
require us to deny sense and reason in physical matters which
are set before our eyes and minds by direct experience or
necessary demonstrations. This must be especially true in those
sciences of which but the faintest trace (and that consisting of
conclusions) is to be found in the Bible.

[...]

If in order to banish the opinion [of Copernicus] in
question from the world it were sufficient to stop the mouth
of a single man – as perhaps those men persuade themselves
who, measuring the minds of others by their own, think it
impossible that this doctrine should be able to continue to find
adherents – then that would be very easily done. But things
stand otherwise. To carry out such a decision it would be
necessary not only to prohibit the book of Copernicus and the
writings of other authors who follow the same opinion, but to
ban the whole science of astronomy. Furthermore, it would be
necessary to forbid men to look at the heavens, in order that
they might not see Mars and Venus sometimes quite near the
earth and sometimes very distant, the variation being so great
that Venus is forty times and Mars sixty times as large at one
time as another. And it would be necessary to prevent Venus
being seen round at one time and forked at another, with very
thin horns; as well as many other sensory observations which
can never be reconciled with the Ptolemaic system in any way,
but are very strong arguments for the Copernican. And to ban
Copernicus now that his doctrine is daily reinforced by many
new observations and by the learned applying themselves to the
reading of his book, after this opinion has been allowed and
tolerated for those many years during which it was less
followed and less confirmed, would seem in my judgement to
be a contravention of truth, and an attempt to hide and suppress
her the more as she revealed herself the more clearly and
plainly.

(home1.gte.net/deleyd/religion/galileo/galileo.html)

Commentary

Clearly religious culture enjoyed hegemony over astronomy in Galileo's time. To defend both himself and science from religious intolerance, Galileo must kowtow to the prevailing religious views. He must speak the language of religion in order to argue in favour of scientific observation and discovery. This is a difficult challenge because of the obvious contrasts between faith in the unseen and unproven and scientific objectivity. To insist on the objectivity of science, he must speak in a language of faith. To save himself from his accusers, Galileo tries to show that he is a believer who accepts the ultimate authority of the Bible. But to save science, he reminds his readers that the Bible's authority extends only as far as its meaning is understood. If the Bible's meaning is unclear, or if the Bible says nothing about those astronomical discoveries made long after the Bible was written, then Galileo concludes, the Bible ought not be the last word on nature's laws. Galileo argues for faith, but it is ultimately faith in 'sensory observations' rather than in the unseen. To ban Copernican theory would mean banning the ability to see; it would be a denial of all that had already been seen in the heavens themselves.

Galileo was not successful in convincing the clergy to accept the Copernican theory of the solar system. Here are excerpts from the church's response to his letter.

> We by the grace of God, cardinals of the Holy Roman Church, Inquisitors General, by the Holy Apostolic see specially deputed, against heretical depravity throughout the whole Christian Republic.
>
> [...]
>
> ... we say, pronounce, sentence, declare, that you, the said Galileo, by reason of the matters adduced in process, and by you confessed as above, have rendered yourself in the judgment of this Holy Office vehemently suspected of heresy, namely, of having believed and held the doctrine – which is false and contrary to the sacred and divine Scriptures – that the sun is the centre of the world and does not move from east to west, and that the earth moves and is not the centre of the world ...
>
> [...]
>
> And in order that this your grave and pernicious error and transgression may not remain altogether unpunished, and that you may be more cautious for the future, and an example to others, that they may abstain from similar delinquencies – we ordain that

the book of the 'Dialogues of Galileo Galilei' be prohibited by public edict.

We condemn you to the formal prison of this Holy Office during our pleasure, and by way of salutary penance, we enjoin that for three years to come, you repeat once a week the seven penitential Psalms.

Reserving to ourselves full liberty to moderate, commute, or take off, in whole or in part, the aforesaid penalties and penance. And so we say, pronounce, sentence, declare, ordain, condemn and reserve, in this and any other better way and form which we can and may lawfully employ.

Economics and science

Economic or political discourse may subordinate science in a number of ways.

The thorny issue of global warming and what industrialized nations ought to do to stem their emissions is one in which science often loses out to economics and politics.

Activity

Text 38 shows excerpts from a letter written by American president, George Bush, in 2001 in which he explains why the US will not sign the Kyoto agreement. In what ways is science subordinated to economic or political interests? What do you notice about Bush's language, his use of the discourse of economics?

Text 38

From a letter from the President to Senators Hagel, Helms, Craig, and Roberts.

For Immediate Release, Office of the Press Secretary, 13 March 2001

A recently released Department of Energy Report, 'Analysis of Strategies for Reducing Multiple Emissions from Power Plants', concluded that including caps on carbon dioxide emissions as part of a multiple emissions strategy would lead to an even more dramatic shift from coal to natural gas for electric power generation and significantly higher electricity prices compared to scenarios in which only sulfur dioxide and nitrogen oxides were reduced.

This is important new information that warrants a reevaluation, especially at a time of rising energy prices and a serious energy shortage. Coal generates more than half of America's electricity supply. At a time when California has already experienced energy shortages, and other Western states are worried about price and availability of energy this summer, we must be very careful not to take actions that could harm consumers. This is especially true given the incomplete state of scientific knowledge of the causes of, and solutions to, global climate change and the lack of commercially available technologies for removing and storing carbon dioxide.

(www.whitehouse.gov/news/releases/2001/03/20010314.html)

Commentary

Certainly the threat of higher energy prices should not be ignored as we deliberate about the best way to protect our global environment for the future. Higher energy costs would lead to higher prices of basic necessities like food, clothing and housing. Yet, Bush, like any politician who wants to use science to back up his position, refers to scientific research as 'incomplete' in order to argue that serious emission controls would be economically disadvantageous. As long as it appears that scientific research is 'incomplete', he is saying, then we don't need to make drastic changes.

He and other politicians who fear economic catastrophe if we force polluters to curtail their emissions get away with labelling scientific research as incomplete because scientific communication about global warming is necessarily cautious and objective. What is 'incomplete' to a motivated politician searching for the language of fact and certainty in scientific communication may, to a scientist, simply mean that mathematical models have been used for predictions and that we won't know for sure until the future is now.

Activity

In Text 39 are shown the first two paragraphs from the summary of a report written by a team of scientists commissioned by the US National Research Council. Assume President Bush has read this report. Why would he label scientific research on global warming 'incomplete' if he based his judgement on these passages?

Text 39

Greenhouse gases *are accumulating* in Earth's atmosphere as a result of human activities, causing surface air temperatures and subsurface ocean temperatures to rise. Temperatures are, in fact, rising. The changes observed over the last several decades *are likely mostly due to human activities*, but we cannot rule out that some significant part of these changes is also a reflection of natural variability. Human-induced warming and associated sea level rises *are expected to* continue through the 21st century. Secondary *effects are suggested by computer model simulations and basic physical reasoning*. These include increases in rainfall rates and increased susceptibility of semi-arid regions to drought. The impacts of these changes *will be critically dependent on* the magnitude of the warming and the rate with which it occurs.

The mid-range model *estimate* of human induced global warming by the Intergovernmental Panel on Climate Change (IPCC) *is based on the premise* that the growth rate of climate forcing agents such as carbon dioxide will accelerate. The *predicted* warming of 3°C (5.4°F) by the end of the 21st century is consistent with *the assumptions* about how clouds and atmospheric relative humidity will react to global warming. This estimate is also *consistent with inferences* about the sensitivity of climate drawn from comparing the sizes of past temperature swings between ice ages and intervening warmer periods with the corresponding changes in the climate forcing. This *predicted temperature increase is sensitive to assumptions* concerning future concentrations of greenhouse gases and aerosols. Hence, national policy decisions made now and in the longer-term future will influence the extent of any damage suffered by vulnerable human populations and ecosystems later in this century. Because there is *considerable uncertainty* in current understanding of how the climate system varies naturally and reacts to emissions of greenhouse gases and aerosols, current esti-mates of the magnitude of future warming should be *regarded as tentative and subject to future adjustments* (either upward or downward).

Commentary

The report urges that 'national policy decisions made now and in the longer-term future will influence the extent of any damage suffered by vulnerable human populations and ecosystems later in this century'. Yet a reader, such as an American president who fears making decisions that will lead to economic sacrifice for manufacturers and voters as well as political disaster for himself may happily 'misread' the scientific discourse here. Words such as 'inference', 'assumption', 'predict', and 'uncertainty' can be easily interpreted to mean that the science of global warming is voodoo. No wonder president Bush called scientific research on global warming 'incomplete' in order to justify not signing the Kyoto agreement.

SUMMARY

Understanding how any scientific field interacts with culture is perhaps just as important as understanding its content and methods. Scientists need to be aware of how cultural discourses influence their own attitudes and behaviours no matter how objective they try to be. Non-scientists need to be aware of the subtle and not so subtle affects of science on their cultural attitudes.

Science and society

We owe our modern lives to the scientific enterprise. That's a grand statement. But if we consider all of those areas of exploration that can be described as 'scientific' – the social sciences as well as the natural sciences – we can applaud those who measure and observe and count and describe their subjects precisely and as objectively as possible. The results of their endeavours have improved our health, built our bridges and cities, taken us into space, increased our food production, enhanced our institutions and communities.

Of course, there is always a downside. The atom bomb, pollution, chemical weapons and the global spread of new viruses can all be attributed, directly or indirectly, to scientific advances. Scientific advances also force us to confront our traditional values. The introduction of birth control pills, the first test-tube baby, the first cloned sheep and the first removal of stem cells from a human embryo are all examples of advances that brought about enormous public discussion. Traditionalists want to preserve what they consider 'God's domain' or what is 'natural', while progressives argue that human intellect should be free to remake our world and that values adjust to new circumstances made possible by science.

Whether the products of science bring about benefits or deficits to our lives, one thing is certain: we pay attention when science speaks. The language of science, that discourse that makes us sit up and take notice, has a powerful impact no matter whether it comes from credible sources or quacks – or some source in between.

Our exposure to scientific knowledge is most often in the form of translations. As we learned in Unit three, the 'congruent' form of a scientific statement is a type of translation that is more easily understood by non-scientists. Unfortunately, as with any translation from one language to another, some meanings can be changed or even lost. Scientific research reports can be difficult, even impossible, for non-scientists to understand. Those who translate them can easily misunderstand or intentionally dramatize or 'misread' their results.

TRANSLATORS OF SCIENCE AND THEIR MOTIVES

Translations of science are motivated by a variety of concerns. Here are some of the translators of science and their motives:

Translator 1: Journalists translate science for the general public

All journalists want a good 'story', something that will attract readers because of its importance, drama or ultimate impact on their lives. Because of this motive, journalists may overstate the findings of a particular report. Journalists with a background in science are more capable of understanding complex data than are untrained journalists. But all journalists are expected to tell a good story which means that they may report on only the most potentially significant scientific research and ignore those areas that seem less important to readers. This means that we simply do not learn of recent research about, for example, a new species of frog or arcane areas of mathematics. Our exposure to science in the media is limited by the expertise and motives of journalists reporting science.

Below is a statement of the main findings from a scientific study, a longitudinal survey (Christakis and Zimmerman, 2004) of parents concerning childhood attention problems:

In a logistic regression model, hours of television viewed per day at both ages 1 and 3 was associated with attentional problems at age 7 (1.09 [1.03–1.15] and 1.09 [1.02–1.16]), respectively (708).

[. . .]

Early television exposure is associated with attentional problems at age 7 (708).

Here we have the nominalizations 'hours of television viewed' and 'early television exposure' that contribute to the claim that TV and attention problems are 'associated'. While the researchers clarify that none of the parents reporting 'attentional problems' have had their children diagnosed with ADD or ADHD, clearly, the theory of future diagnoses is already developing. With television viewing or exposure as the 'thing' or agent, other factors such as genetics or the possibility that kids with attention disorders simply watch more TV drop out as 'things' or agents in these processes (even though the researchers add these disclaimers in their discussion sections).

Activity

Text 40 is an excerpt from a journalist's account of the scientific study. How does the journalistic account differ from the original?

Text 40

The more television children watch between the ages of 1 and 3, the greater their risk of having attention problems at age 7, researchers reported on Monday. They found that each hour of television that preschoolers watched per day increased the risk of attention problems such as attention deficit-hyperactivity disorder, by almost 10 percent later on.

(Conion, 2004)

Commentary

The language changes may be charted as follows:

Technical	→	Congruent
'Television exposure'	→	'watch'
'Associated with'	→	'increases the risk of'
attention problems	→	attention deficit-hyperactivity disorder

The technical style is vulnerable to journalistic rewriting, possibly leading to overstated or misleading reports. The vagueness of nouns, 'exposure', 'associated with' and 'attention problems', invites reconstrual into verbs 'tied to' (Conion) or 'increases the risk of' (Conion) and specifying 'attention problems' as ADHD. Since we are all concerned about the amount and type of television children watch, the congruent rewritings of this study may do more good than harm. But other rewritings, motivated by different ideologies or interests, may cause more harm than good.

Translator 2: Public officials often translate science in making or promoting public policy

Public health officials, government agency representatives and officials, and leaders of various institutions all convey scientific information for particular purposes. Depending upon their purpose, officials may emphasize only the 'good news' or only the 'bad news' from scientific research.

Activity

The passages in Texts 41 and 42 contain public health statements about the SARS epidemic. Examine them for their purpose. Do they emphasize the good news or the bad news and why?

Text 41

From the SARS Action plan of the Asian Development Bank, May, 2003 (www.newton.uor.edu/Departments&Programs/AsianStudiesDept/sars.html).

The longer the disease lasts – and major medical questions remain unanswered – the more serious the economic and social impacts will be. It is not clear at this time how this epidemic will play out. A recent paper proposes possible scenarios that may prevail. Given the continuing uncertainty about SARS nature and transmission, the actual outcome could be within this range, or could be worse.

Regardless of which scenario prevails, SARS is already having a major economic, social, and psychological impact on the populations of the countries most affected by it. Some of the impacts are already visible, while others are likely to be felt over time, depending on the duration of the epidemic.

Text 42

From a case report of Congressional-executive Commission on China, 'Information Control and Self-Censorship in the PRC and the Spread of SARS' (www.cecc.gov/pages/news/prcControl_SARS. php?PHPSESSID=adb4443bfdc222c4851eef7e6b244407)

According to reports in PRC government newspapers, health care workers in Guangdong province began noticing people coming in with 'atypical pneumonia' in mid-November, and by early January people were already engaged in panic buying at drug stores because of rumors of the spread of a 'mystery epidemic'. But the same government-controlled newspapers that first broke the story in early January devoted most of their coverage to stories with headlines telling people that 'The Appearance of an Unknown Virus in He Yuan is a Rumor' and articles quoting government claims that 'there is no epidemic' and that illnesses were 'the result of changes in the weather leading to a decline in people's immune systems. The people of the PRC (and the world) would not learn the truth until the disease began killing people in Hong Kong, where government restraints on the free flow of information are not as severe as in mainland China. Only then would the PRC's central government acknowledge the disease's existence. Even then, the government-controlled media continued to insist that everything was under control.

Commentary

This case report states that the initial response to SARS by Chinese government officials was to silence the 'bad news' of a possible epidemic disease. Fear of inducing panic among citizens, travellers and international businesses dealing in China led bureaucrats to dismiss the very real threat of SARS, leading to a public health disaster and even worse consequences for citizens, travellers and business than would have been the case if they had acknowledged the threat directly and openly.

Representatives of governmental agencies may also be motivated to represent science in particular ways. In the United States, those who promote the use of fossil fuels speak a very different scientific discourse than those who promote conservation.

Activity

Compare the statements shown in Texts 43 and 44.

In Text 43 the United States Assistant Secretary of Energy for Fossil Energy looks at when best to drill in the Arctic tundra (the Department of Energy's study on Drilling in the Alaskan Wildlife Refuge). Text 44 is an extract from a report by the US National Academy of Science concerning damage to tundra from the off-road travel necessary for oil exploration.

How do these two groups use scientific discourse to justify their conclusions? Which one seems more credible and why?

Text 43

Sound science offers the best way to protect sensitive environments. Today, however, all we have is a general 'rule of thumb' for determining when it is environmentally safe to move oil exploration equipment across the Arctic tundra. This project will apply the latest scientific instrumentation and modeling to refine our understanding of the tundra's resistance to disturbances. The result will be better environmental protection and a much more scientific basis for determining when oil operations can be conducted.

(US Department of Energy,
Office of Fossil Energy)

> Surface erosion, water flow and tundra vegetation on the North Slope have been altered by extensive off-road travel. Some damage has persisted for decades. The current 3-dimensional survey method requires a high density of seismic-exploration trails. Networks of these trails now cover extensive areas and are readily visible from the air, degrading visual experiences of the North Slope. Despite technological improvements and increased care taken by operators, the potential for damage to the tundra still exists because of the large number of vehicles and camps used for exploration.
>
> (National Academies Press online,
> www.nap.edu/books/0309087376/html/)

(Note: there is no commentary for this activity.)

Translator 3: Advertisers exploit scientific discourse to sell everything from mattresses to drugs

Since advertisers are motivated by profits, they can be the least credible translators of scientific information. The combination of the discourse that makes us sit up and take notice and the profit motive can be disastrous.

A historical example is cigarette advertising in the United States in the 1940s and 1950s. The health benefits of smoking were promoted in television and magazine adverts, often with actors in white coats.

We may all be able to detect the falsity of a character in an advert who says, 'I'm a research scientist and I'm here to tell you about a new breakthrough in cleaning products.' But we may not be as apt to distinguish credible from credible sources in more subtle adverts.

The marketing of prescription drugs typically provides some baseline information about a drug while espousing its benefits. For the general public, drug adverts emphasize the 'good news' of the drug's benefits, usually through dramatic images of happy and healthy people. The 'bad news', usually the side effects and the list of people who should not take the drugs, is typically relegated to a less noticeable space on the page or to a monotone voice-over of those side effects if a television advert.

General audiences for these adverts are quite vulnerable to the hopeful message they provide. One of the most vulnerable audiences for drug adverts are those who have the virus that causes AIDS, HIV. In the early 1990s, new drug cocktails, combinations of drugs that inhibit virus replication, were introduced into the market. The companies selling these drugs began to market not only to doctors but also to those who were HIV positive. Several adverts produced by one American drug company, Merck, appeared in a magazine for people who are HIV positive, on bus shelters, billboards and other public spaces.

The adverts depicted muscular, healthy-looking men participating in vigorous activities with the captions 'Going the Distance' and 'Living with AIDS'. The sales message to consumers was that a miracle cure had been discovered, that by taking a pill, the HIV-positive patient or the AIDS patient would not only be able to live with AIDS but do so like an Olympic athlete.

The adverts portrayed only part of the story, unfortunately. Though the new drugs were very promising – in the early drug trials patients responded well – there were and still are many problems with these drugs. They do not cure people, nor do they work in all patients. Some patients cannot tolerate the drugs, and some mutations of the virus are not affected. The drug adverts' seductive images created false hope for many people.

Such adverts were also damaging to efforts to stem the transmission of the AIDS virus. Adverts featuring athletic and handsome actors may have instilled a more carefree attitude toward the disease, leading to unsafe sexual practices. Adverts that depict AIDS as an inconvenience or a survivable disease mislead consumers.

Eventually the United States Food and Drug Administration ordered the adverts changed, criticizing drug companies for misrepresenting patients' lifestyles and the drugs' abilities to improve health of HIV-positive people.

Translator 4: Scientists themselves are often the most significant translators of science for society

We are all grateful to scientists such as Stephen Hawking and Carl Sagan who can translate the most intricate and complex ideas in a compelling and understandable manner. What ensures a good translation, one that readers will enjoy and understand?

The use of metaphor that we easily recognize to illustrate or explain difficult concepts is one strategy of a good translation. As we learned in

Unit one, metaphors exploit some similarity between two very different things. When we use metaphor to help us translate or explain science, we must be aware that our metaphors can emphasize some qualities but not all qualities of the thing we are explaining. A well-chosen metaphor, though limited in its explanatory capacity, must not mislead or oversimplify.

Another strategy is to communicate more intimately and emotionally with readers than scientists do with each other. A tone of voice that is rich with excitement and curiosity, energy and personality can create an intimate connection with readers. Using active voice rather than passive voice, the second person (you and we) rather than the third (they and he/she) and asking questions will eliminate the dense prose that readers find difficult. Humour, even cartoons, can also enliven a translation written for the general reader.

Activity

Text 45 shows an excerpt from mathematician and physicist Brian Greene's *The Elegant Universe*. What qualities in this passage do you think were intended to help the general reader understand the strength of electro-magnetism.

Text 45

Imagine holding an electron in your left hand and another electron in your right hand and bringing these two identical electrically charged particles closer together. Their mutual gravitational attraction will favor their getting closer while their electromagnetic repulsion will try and drive them apart. Which is stronger? There is no contest: The electromagnetic repulsion is about a million billion billion billion billion (10^{42}) times stronger! If your right bicep represents the strength of the gravitational force, then your left bicep would have to extend beyond the edge of the known universe to represent the strength of the electromagnetic force.

(Greene, 1999: 12)

(Note: there is no commentary for this activity.)

Translator 5: The quack or pseudo-scientist

One false or pseudo science is **Eugenics**, and it continues to force its way into public discussion. Eugenicists believe that human intelligence is largely hereditary, that civilization depends totally upon innate intelligence, that the human race is becoming less intelligent with every generation because the less intelligent are having the most children. They believe in genetic engineering and selective breeding as measures to improve the human gene pool and thus improve civilization. The reason Eugenics is not a credible science is because it is based on several faulty assumptions. Eugenics has been, to different degrees, a matter of public dialogue and public policy since Charles Darwin introduced his theory of the origin of the species. During the height of the Eugenics Movement in Great Britain and in the United States, compulsory sterilization of both men and women who exhibited physical or mental abnormalities or low intelligence was standard. Certainly institutionalized individuals, such as criminals and the mentally ill, were sterilized. Hitler was perhaps the most famous eugenicist with his desire to build a better Aryan race and eliminate the 'inferior Jew'. Eugenicists and racists are close cousins. Still, the eugenicists continue to make their claims.

SUMMARY

We should all be aware that most of the scientific ideas we read or hear about are translations. Each translation is motivated by various concerns whether they are to tell a dramatic story to gain readers or to protect agendas or to promote public policy or to make money. Our critical attention to the goals behind these translations will help us participate in the legitimate aims of public science and protect us from the illegitimate designs of fakes and quacks.

Extension

1 Drug advertising: Find out how drug advertising is controlled and monitored in one country. Have there been any examples of misleading and damaging drug advertising?

2 Translating science: Visit the section of your library containing recent scientific and medical journals. Find a research report on a topic that interests you. You may also look up a topic in an internet database available in your library. Read the article, paying special attention to the abstract and the discussion sections because these will provide the main conclusion of the study. Write at least one translation for non-scientists. Try to use some of the strategies for making an explanation interesting and understandable to general audiences.

3 Public health: Research a public health emergency in your own region. Find news reports that quote public health or government officials. How do they respond to this threat?

4 Scientific reports: Look up a report in some news outlet of some scientific or medical discovery. Take from the report any names of the scientists responsible for the research. Next, look up the scientists' names in the Web of Science database or in any general science database available to you. Once you find the scientific report that provided the basis for the journalistic report, compare the two. Do you find the 'rewriting' of scientific discourse into popular discourse? What changes in meaning and even misunderstandings are produced by these rewritings?

FURTHER READING

Greene, Brian, *The Elegant Universe: Superstrings, Hidden Dimensions, and the Quest for the Ultimate Theory*, New York: W.W. Norton, 1999.

SOURCES

Christakis, Dmitri A. MD MPH and Zimmerman, Frederick J. PhD, 'Early Television Exposure and Subsequent Attentional Problems in Children', *Pediatrics* 113.4 (2004): 708–14.

Conion, Michael, 'Toddler TV Habits Tied to Attention Deficit Study', MSNBC Wire Services, 6 April 2004: www.reutershealth.com/en/index.html.

index of terms

AIDS 14
Acronym for 'Acquired Immunodeficiency Syndrome,' a disease caused by the Human Immunodeficiency Virus (**HIV**). A person may be infected with HIV for many years before becoming severely immune deficient. Infected individuals who do not realize they are infected can easily infect others through exchange of body fluids through sex and needle sharing.

analogy 17
A comparison between two things that share certain features or between a part of one thing and another thing. Evolutionary biologists use the term to indicate functional similarity between organisms that have developed similar features because of a common environment rather than descent from a common ancestor.

bias 1
A conscious or unconscious viewpoint or process that leads to a deviation from verifiable knowledge. In the media, journalists may hold views that lead to a deviation from accurate news reporting. In science, the process of setting up a study may be biased if study subjects are selected in a particular way rather than randomly. Results of such studies do not contribute to knowledge.

black-boxed 95
An expression coined by Bruno Latour and Steve Woolgar to refer to the point at which an issue is no longer debated in science and often difficult to challenge.

clinical discourse 15
The patterns of language and communication that have evolved within the clinic where disease is studied and treated. One may observe the evolution of this language simply by reading clinical reports in the *The Lancet* or *The New England Journal of Medicine* over the last one hundred years.

conjecture 60
A plausible guess that serves as a tool in solving problems. Unlike a **theory** or a **hypothesis**, conjecture is a first step in imagining how a problem might be solved or what the answer might look like. The imaginary solution or answer can be tested against observations.

connotation 10
An implied meaning of a word or a meaning of a word derived from associations, such as cultural or political viewpoints. Scientists try to eliminate terms that carry misleading connotations.

congruent 38
In the **theory** of **grammatical metaphor**, a first step in making

sense of our experience through language. Our experience enters our language as action, as in, 'Yesterday, I read a novel for pleasure'. This is a congruent expression. But we often move beyond the experience of action to the experience of theorizing. Our language supports our theories because it allows us to change a congruent expression into a theoretical one, as in 'Reading novels is a pleasant way to spend the day'. **Scientific discourse** often moves away from the congruent into the theoretical.

data 1

The plural of datum. Data are the collected and measured factors, characteristics, or attributes of individuals or systems. In scientific experiments, conditions may be altered or controlled in order to account for a variety of environmental factors so that the data collected are more reliable.

distilling 43

In the **theory** of **grammatical metaphor**, the recasting of a scientific term into various parts of speech.

DNA 17

Deoxyribonucleic acid; DNA molecules encode hereditary information and pass it from generation to generation.

Eugenics 124

A social movement that advocates the improvement of human beings through various **methods** that include selective breeding, sterilization, genetic engineering, and extermination. Eugenicists

assume that our genes play a stronger role than our environment in shaping our intelligence and behaviour and that these traits are heritable.

evidence 3

What constitutes support for a conclusion or claim or **theory**. Scientists making claims about the natural world must provide evidence that has been gathered according to sound scientific **methods** of inquiry. Evidence comes in many forms. For example, evidence can be physical such as the fossil record as evidence for evolution or it may be clinical such as the outward signs and symptoms of particular processes or diseases. Evidence can also be the repeated observations that lead to generalizations about the natural world.

experimental report 50

An account of laboratory research that follows the scientific **method**: the problem is defined, a **hypothesis** is created, experiments are devised to test the hypothesis, experiments are conducted, and conclusions are drawn. Reports go through a peer review process to ensure quality of **methods** and results.

fact status 59

The condition of consensus that a claim or conclusion reaches after collective experience and observation have found it reasonable and/or verifiable.

first-person pronouns 49

'I' OR 'We' in English.

genetics 17

The study of heredity.

grammatical metaphor 38

The shifting from actions to things in language that allows us to theorize about future experience. While we may have experienced an action, such as 'I drank water out of the stream', we can metaphorically shift from that action to this statement: 'Drinking water out of streams is safe.' This is part of the process of making predictions that our language encourages.

hedge 59

As a verb, to hedge is to indicate the level of certainty in a statement, as in 'X might be the cause of Y'. As a noun, a hedge lessens the certainty of an observation.

HIV 16

Human Immunodeficiency Virus: a **retrovirus** that causes the infection that leads to **AIDS**.

homology 16

Detailed similarity between organisms that is not due to a similar response to a common environment or to similarity in function. Homology is similarity between organisms due to descent from a common ancestor.

human genome 33

The full **DNA** sequence of a human being; all the genes in the human chromosomes.

hypothesis 25

An explanation of an object or process in nature or a prediction of the existence of objects or processes in nature. Scientists test their hypotheses through experimentation or other forms of investigation.

immunology 24

The study of the immune response or immunity.

lexical metaphor 39

The verbal representation or explanation of one domain of experience through reference to another domain of experience.

linguistic transformation 39

The transforming of experience through language; our experiences are not only captured and presented by our language but actually changed by our language.

metaphor 3

A comparison between categorically different objects, processes, or experiences. Metaphors can be **lexical** and **grammatical**. They can also be **root**ed in our way of ordering our world.

method 7

A series of steps followed in order to arrive at an outcome; in science methods of observation, investigation, experimentation, calculation, etc. are indispensable to the growth of knowledge.

modalities 61

Terms indicating conditions needed to make a statement or claim plausible. Modalities include references to human agents, to the time of discovery, or to the circumstances or conditions. These statements may also contain qualifiers such as 'high', 'low', 'novel', or any other descriptor that increases or decreases the

laboratory replication; agreements that a scientific community has reached about the nature of some phenomenon.

scientific rhetoric 3

The art of using the resources of scientific language and argument to convey a conclusion to an audience.

subjective 7

Based on personal feeling or opinion; not **objective**.

terminology 3

Words or phrases used to describe or explain phenomena.

theory 18

In science, an explanation that can be tested by experience in the laboratory or in the field. Mature scientific theories organize knowledge and explain and predict objects, events or processes.